Managing technological discontinuities

This volume draws on the latest research to provide an innovative analysis of the processes through which managerial actors implement technological discontinuities at firm level. The conceptual framework of management mobilisation and technological change is illustrated by original empirical evidence on the Finnish paper industry sector, which has achieved a strong strategic position in its field and been at the forefront of significant technological innovation. The in-depth longitudinal analysis examines the relationship between managerial action and discontinuous technological change.

Managing Technological Discontinuities is a rich and theoretically grounded analysis of the processes of change which provides a corrective to many under-theorised accounts of technological change and the management of change.

Dr Juha Laurila is a senior lecturer at the Helsinki School of Economics, Finland. He has recently finished a three-year research project on management of technology funded by the Academy of Finland. His recent publications include articles in such scholarly journals as *Journal of Management Studies*, *The International Journal of Human Resource Management* and *Scandinavian Journal of Management*.

The Management of Technology and Innovation

Edited by David Preece
University of Brighton/SPRU, UK

The books in this series offer grounding in central elements of the management of technology and innovation. Each title will explain, develop and critically explore issues and concepts in a particular aspect of the management of technology/innovation, combining a review of the current state of knowledge with the presentation and discussion of primary material not previously published.

Each title is designed to be user-friendly with an international orientation and key introductions and summaries.

New titles in this series include:

Teleworking: International Perspectives

From Telecommuting to the Virtual Organisation
*Edited by Paul J. Jackson, Brunel University, UK, and
Jos M. van der Wielen, Tilburg University, The Netherlands*

Managing technological discontinuities

The case of the Finnish paper industry

Juha Laurila

Routledge
Taylor & Francis Group

LONDON AND NEW YORK

First published 1998 by Routledge
2 Park Square, Milton Park, Abingdon, Oxon OX14 4RN

Simultaneously published in the USA and Canada
by Routledge
605 Third Avenue, New York, NY 10017

Routledge is an imprint of the Taylor & Francis Group, an informa business

© 1998 Juha Laurila

Typeset in Bembo by Routledge

British Library Cataloguing in Publication Data
A catalogue record for this book is available from the British Library

Library of Congress Cataloging in Publication Data
Laurila, Juha.
    Managing technological discontinuities : the case of the Finnish
    paper industry / Juha Laurila.
    p. cm.
    Includes bibliographical references.
    1. Paper industry–Finland–Technological innovations.
    2. Technological innovations–Management. I. Title.
    HD9835.F5L38 1998   98–3179
    676'.2' 0685–dc21   CIP

ISBN 13: 978-0-415-17853-2 (hbk)

For Eija and Jarkko

Contents

Illustrations

Preface

In this section I will briefly describe the distinctive characteristics of this volume in relation to previous literature on the field. I will also give my views on its intended audience. Much of the usual prefatory comment, however, is contained in the Introduction.

This book addresses problems related to the management of technological change. The topic is not new in the sense that there are already a number of books with an organisational and management perspective on technological change and innovation. However, it seems that there is still room for another volume in this domain. This is, first, because most of the previous books on this field are descriptive rather than explanatory in focus. In other words, although we have several volumes prescribing how to implement technological change, we lack systematic analyses of the managerial processes related to this implementation. Second, remarkably few of the present volumes undertake in-depth longitudinal analyses on the relationship between managerial action and discontinuous technological change. In general, most of the existing literature has considered technological discontinuities an external condition which forces management to make adaptive responses. In contrast, the firm-level managerial action which both generates and aims to overcome these discontinuities has received much less attention. The present book confirms that an adequate conceptualisation of these processes would require adoption of a longitudinal research approach and would benefit – at least at this stage – from an idiographic research design. Third, few if any of the previous contributions on this field systematically develop a framework for analysing the mobilisation of managers to initiate and implement technological change as a contextually determined phenomenon with consecutive upswings and downturns. Instead, management is most often treated as a monolithic whole whose action is restricted only by cognitive limitations. By demonstrating how management consists of conflicting actors and how this constellation of actors may become temporarily united and again divided as a consequence of temporary resource changes and changes in the corporate competitive environment, this book offers an innovative approach to the analysis of organisation and technology.

This book introduces a conceptual framework to analyse how technological discontinuities are managed. The framework includes categories related to both managerial actors and technology. It is illustrated with original empirical evidence on the Finnish paper industry sector which has achieved a strong strategic position in its field in Europe and worldwide. For example, Finnish paper industry companies have the most efficient and advanced production facilities and are therefore competitive suppliers of paper industry products across the globe. This means that the present book has an empirical focus on an industry which has, despite its geographically peripheral location, reached a prominent position in its business. I personally believe that the relevance of the book for an international public is increased by the fact that Finnish paper industry firms are the main technological innovators in their field in Europe and worldwide. By demonstrating how technological discontinuities are managed in such contexts the book offers important insights for the management of technology in general.

I expect that this book is of interest to a wide range of researchers and students. In general, this is because it combines conceptual models of management and technological change with longitudinal empirical evidence and illustration. In particular, this is because the book significantly extends previous conceptualisations on the mobilisation of managerial actors to initiate and implement technological discontinuities. The book can be used both by readers specialising in these issues and also as a supplementary text on courses in the technology and management domain. The in-depth analyses of discontinuous technological change and managerial action are relevant material for understanding the behavioural requirements of discontinuous change that are frequently mentioned but rarely studied thoroughly. The book is also relevant as supplementary material on courses concerning industries in contemporary Europe. Although the main audience is involved in the field of management and organisation studies, I believe that the book will also be of interest to all readers following the development of industries within the European Community. Moreover, I believe that the book's longitudinal analyses of organisational and technological change are appropriate to a wider audience including MBA students and practitioners in fields such as organisational behaviour and organisational change. For example, the processual perspective adopted to technological change will find interested readers among managers and scholars concerned with issues such as the design and implementation of innovation projects.

Acknowledgements

This book is the outcome of empirical research made between 1989 and 1997. It is therefore the result of a research effort for which I received assistance from a number of persons and institutions. I would therefore like to mention at least some of those who contributed to this endeavour.

I would like first to thank all those individual managers who acted as informants for the research reported here. This study would not have been possible without the co-operation of the numerous managers interviewed and the managers who have commented on my findings and interpretations in oral or written form. Here I would also like to thank Esa-Jukka Käär from the Finnish Forest Industries' Federation for providing updated information on the paper industry.

I would like to thank the various colleagues who have shared their ideas with me during the various phases of this project. During my doctoral studies between 1990 and 1994 I benefited from the advice and supervision of three senior scholars at the Helsinki School of Economics: Kari Lilja, Keijo Räsänen and Risto Tainio. They all substantially influenced my work by arranging research facilities, commenting on drafts and sketching fruitful paths for further study in this field. I should really single out Kari here; he originally encouraged me to do research on the paper industry and has provided the necessary support and criticism whenever needed. Other important Finnish colleagues from whom I have learned include Päivi Eriksson, Sirkku Kivisaari, Kimmo Kuitunen, Mauri Laukkanen, Matti Pulkkinen and Arja Ropo.

This work is also part of the emerging research on paper industry management. In developing my research agenda in this field I have benefited from discussions with a group of researchers including Kimmo Alajoutsijärvi, Marja Eriksson, Katri Gyursanszky, Mika Huolman, Jan Jörgensen, Markku Kuisma, Anneli Leppänen, David MacGregor, Roger Penn, Jukka Ranta, Carl-Johan Rosenbröijer and Thomas Rohweder. Other scholars, whom I wish to thank for commenting on my work without taking any responsibilities for the ideas presented here include Robert Burgelman, Margaret Grieco, Dick Scott, Haridimos Tsoukas and Richard Whitley.

My thanks go to the Routledge Series editors David Preece and John Bessant for taking the original book proposal seriously. I especially value the comments and encouragement by David at that stage. At the later stages of the project the commissioning editor Stuart Hay and his assistant Craig Fowlie were of considerable help. As part of the publishing process, I received comments on the original book proposal from four anonymous Routledge reviewers for whom I also would like to express my gratitude. When preparing the manuscript for this book Sampo Tukiainen helped me with both bibliography and with references and illustrations. I am also most grateful to David Miller for the work he has done in improving the language of this book.

Acknowledgement is given to a range of institutions for funding the preparation of this book. These include The Academy of Finland and the Helsinki School of Economics and Business Administration and its foundations. I have also received additional funding for this project from the Marcus Wallenberg Foundation.

A small part of the material in this book has already appeared as progress reports of a kind. I would like to thank the publishers of those papers for permission to adapt and republish those parts that have been used. Seven direct quotations have been adopted from S. Sackmann (ed.) *Cultural Complexity in Organizations* (pp. 252–272, copyright © 1997 by Sage Publications Inc.) and reprinted by permission. Figure 5.2 and Tables 4.2 and 4.3 have been adopted from the *Journal of Management Studies* (Vol. 34, pp. 219–239, copyright © 1997 by Blackwell Publishers Ltd.) and reprinted by permission.

Finally, I need to confess that my wife Eija was again the critical supporting person behind this work. She has for many years now shared both her emotions and intellect with me and deliberately worked to make our life enjoyable. I dedicate this book to her and to our son Jarkko: our most demanding and gratifying challenge so far.

Juha Laurila
Helsinki, Finland
January 1998

1 Introduction

Some theoretical background

The relationship between management and technological change is a common theme in previous management and organisation literature. This existing body of work demonstrates conclusively that management both enables and restrains changes in the technological core of firms and companies. This is because although managers have power to introduce new technology, they have cognitive and behavioural limitations which often inhibit initiation of technological change. However, as some seminal works within management and organisation literature (e.g. Child 1972) clearly show, studies of technological change should pay special attention to managerial actors because corporate structures in themselves are incapable of producing change. In this spirit, we may assume that the technological development of business firms is influenced most by the characteristics and conduct of their key managers (e.g. Hambrick and Mason 1984; Tushman and Romanelli 1985).

However, assuming that managers make a difference does not imply that the nature of their impact on the technological development of firms is easily anticipated. In contrast, management influence on technological change is in itself a complicated phenomenon. This is both because of the changing cognitions of the managerial actors and the fact that instead of being a harmonious whole, management is divided into groupings with conflicting interests. The research reported in this book addresses the questions that have been emerging since the identification of the political (e.g. Pettigrew 1973) and informal (e.g. Dalton 1959) nature of managerial action. This means that we assume that management includes several actors whose interests do not necessarily coincide (cf. e.g. Noon 1994: 31; Webb and Dawson 1991: 203). Management and managerial actors in large corporations are characterised by diverse internal divisions and micro-political conflicts, partly because of the distinctive and in some sense conflicting objectives of actors on different management levels and functions (Teulings 1986). Moreover, managers in a business firm have different experience bases and represent various professions and managerial subcultures (e.g. Gouldner 1954; Crozier 1964). Therefore managerial actors make divergent interpretations (Weick 1979) of the corporate environment, resulting in support for distinctive strategic action.

This study focuses on the new research problem arising from the realisation that management – like any other social actor – is divided. In other words, how does a subdivided management cope with discontinuous technological change that necessitates at least temporary consensus among organisational actors. In the main, such technological discontinuities have previously been examined on the industry level (e.g. Tushman and Anderson 1986; Anderson and Tushman 1990). The finding in these studies is that discontinuities (e.g. revolutionary technological innovations) make previously competent firms technologically obsolete and give way to the emergence of new ones. Although these are major contributions to our understanding of technological change in general and how technological change influences development of industries and populations of firms in particular, they are nevertheless limited in relation to the issues addressed in this book. This is because they emphasise the technological environment of firms and consequently neglect the managerial processes within them.

In contrast, the problems addressed in this study are related especially to what happens within the firm when the overall technological development is converted into radical change projects and the creation of new businesses. Technological discontinuity is therefore defined here as a situation in which the product or production technology of the firm is altered in a way that makes the previous capabilities obsolete and requires a rapid adoption of new ones on the different organisational levels. From this point of view, the replacement of old technologies is a discontinuous event which requires both adoption of new skills and behaviours and abandonment of previous ones. The replacements of previous technologies are especially critical in capital-intensive industries, in which the traditions in a specific line of production may be long and the financial and social costs of adopting new technologies are high. For example, this book highlights the discontinuities involved in moving a company with a lengthy tradition in producing technologically simple and low value-added paper product (newsprint) to the production of technologically demanding and highly value-added paper product (coated magazine papers). In practice, an old but still functioning paper machine with its low grade paper product was dismantled and a new state-of-the-art paper machine of a sophisticated paper grade with more than double the output was built in its place. The discontinuity thus implied a break with fifty years of corporate paper-making history.

Technological discontinuities are both technically and socially demanding in the sense that the people involved should be able to give up their previous ways and co-operate to obtain the new skills required by the new technology. This is especially so when, as in the case mentioned above, the company's previous experiences are insufficient in relation to the demands of the new technology. Such abrupt alterations in the line of production can therefore be considered major collective efforts comprising technological uncertainties and significant interruptions to the steady flow of operations. Consequently, it can be expected that management and managerial actors play a critical role when such technological discontinuities are initiated and implemented in firms.

The relevance of the mobilisation of managerial actors for the successful accomplishment of technological discontinuities in firms has been acknowledged in the previous management literature. For example, it has been shown that technological discontinuities within firms cannot be efficiently executed without at least the momentary shared commitment of all the various change agents in the corporation (Kanter 1983). This is because a single key executive or even a dominant coalition in management cannot authorise change that depends on the co-operation of managers representing different levels and functions in the managerial hierarchy. In other words, to bring about technological discontinuities, various managerial factions must be mobilised to support the change. Such consensus is required partly because managerial actors need to create cohesion and commitment to ambitious change objectives to solicit the resources necessary for change (Burgelman and Sayles 1986). Moreover, an agreement about the content and timing of such a discontinuity is not enough, because successful execution and implementation of change necessitates building many kinds of internal and external alliances and coalitions (March 1962). For example, if a company lacks specific skills for operating the new technology, new managerial capacity must be acquired. As will be shown later in this volume, such executive migration may make the building of sufficient consensus within management even more difficult as the diversity within management increases.

Although the issues of management mobilisation and discontinuous technological change have been introduced in the previous literature, we believe that further research should be conducted on these issues. In other words, to clarify how such discontinuities come about in firms we need to extend these conceptualisations. This is especially because the previous conceptualisations do not adequately handle the variety of divisions among managerial actors and how this influences the process of mobilisation. Moreover, we need further empirical investigation on how technological discontinuities are initiated and implemented in firms, for example, because it is not quite clear where managers find the courage and sufficient agreement to formulate and set such endeavours into motion.

The present book therefore further elaborates upon these processes of social mobilisation to discover how this kind of pluralistic and subdivided management overcomes the challenge of technological discontinuities. For example, the book will show that conflicting managerial actors may find a temporary basis for collective action as the opportunities offered by a technological discontinuity reconcile existing managerial subdivisions. In these situations the managers may support the same action though their motives to act are different (cf. Hosking and Fineman 1990: 593–594). Moreover, the book also shows how simultaneous increases in corporate resources and changes in corporate business environment encourage managerial actors to initiate ambitious technological change. The way these issues were approached empirically will be briefly outlined below.

Research design

This study examines the management of technological discontinuities in the context of the Finnish paper industry. We believe that research on technological discontinuities at this stage will benefit from a study focused on a specific setting. There are at least two reasons for this. First, it seems that the previous frameworks are too abstract and insensitive to capture management in its industry- and sector-specific forms (Pavitt 1984; Child 1988). Thus it is believed that only by taking account of the specificities of the business context is it possible to advance our existing concepts and to generate 'grounded theories' (Glaser and Strauss 1967) on this field. This is especially so if the aim is not to explain variation in the patterns of technological change but instead to elaborate on the managerial process through which technological change comes about (see e.g. Burgelman 1996; Mohr 1982). Second, it is believed that the management of technological discontinuities can be most easily observed in settings in which they are typical and distinctively well developed (cf. Yin 1989). The Finnish paper industry is therefore relevant here because the systematical initiation and successful implementation of technological discontinuities has made it technologically the most advanced on its field. Consequently, the examination of technological discontinuities in the context of the Finnish paper industry leads us to results that could not be achieved with a less idiographic (Tsoukas 1989) research approach.

The recurrent discontinuities typical of its long-term technological development therefore make the paper industry companies an interesting object for studying how these discontinuities are managed. This is especially because the technological transformation experienced by the Finnish paper industry companies in recent decades can be considered a long chain of successive technological discontinuities embodied in major technological change projects. As an outcome of these projects, the products and production technologies (see e.g. Rohweder 1993) of the industry have been significantly upgraded. Simultaneously, the scale of production and the sophistication of the production equipment have significantly increased (see e.g. Lilja *et al*. 1992). This means that the successful accomplishment of technological discontinuities in firms leads to accumulation of company assets and competencies. This is particularly so in the case of the Finnish paper industry companies which have 'made history' on the level of the global paper industry. However, to understand how the conflicting managerial actors within the paper industry firms have been mobilised to support the adoption of specific new technologies, a closer look at a specific case at a specific moment in time is needed. This study therefore examines the management of technological discontinuities in the Finnish paper industry management on a longitudinal single-case design.

The reason for adopting a firm-in-sector approach (Whipp and Clark 1986; Child and Smith 1987) in a study of managing technological discontinuities is that it permits capturing an ambiguous phenomenon in a longitudinal setting while taking its sectoral context into account. We need such an approach because the mobilisation of managerial actors for technological discontinuities cannot be conceptualised on a cross-sectional basis. In addition, the single-case study design permits picking up a 'revelatory case' (Yin 1989) which is particularly interesting among a larger group of firms. An in-depth study of such a developed case is especially fruitful in making analytical generalisations (Eisenhardt 1989; Dyer and Wilkins 1991) about the issues in question.

Accordingly, the Finnish paper industry company Tampella has been chosen as the object of this single case study.[1] The reason for choosing this case is that it includes ordinary elements in an exceptional setting. More concretely, Tampella is a typical Finnish paper industry company in the sense that in recent decades it has, like its domestic competitors, accomplished several major technological change projects to upgrade its production facilities. Most often, these discontinuities involve replacement of previous technologies with new significantly more advanced ones or the building of completely new production facilities. Tampella is also a typical Finnish paper industry company because the discontinuity in this study closely resembled those accomplished a little earlier by many of Tampella's domestic competitors.

Tampella was exceptional in the sense that while it was technology-driven, it was significantly less so than most of its competitors. More concretely, although Tampella had accomplished several projects, it had relatively little experience of state-of-the-art technology, especially at its paper mill, which had long suffered from lack of investment. Consequently, although the discontinuity studied in depth in this book is not exceptional, the case is deviant because Tampella can be considered a long-term technological 'laggard' among Finnish paper industry companies. The case of Tampella is therefore especially interesting for the study on how technological discontinuities are managed because a company with relatively little previous experience faces the greatest managerial challenge when it tries to catch up with its competitors through a major technological change project. This project implied adoption of technologically advanced and demanding production equipment and dismantling of the existing machinery on exceptionally short notice. The opportunity to observe how the managerial actors at the different levels of the managerial hierarchy were mobilised to initiate and implement this ambitious discontinuity and thus to fill the huge action gap between the requirements of the new technology and the lack of previous experience makes Tampella a revelatory case in its sector.

However, an interesting business context, and a firm, does not carry us far without adequate research access. Access to a large variety of research material is particularly critical to a single-case study. From this perspective, Tampella offered exceptionally good opportunities for field work. The author was allowed considerable freedom to make contacts with actors on different levels of the managerial hierarchy. This is not to say that formal acceptance guaranteed the co-operation and rapport of individual informants (see e.g. Laurila 1997c). However, the case of Tampella was also a good choice in this sense because it offered both a large number of external and internal documents and contact with managers seemingly interested in co-operation with the researcher. This was a result of the dramatic events that took place at Tampella shortly before and during the research process. The problems involved in the technological discontinuity seem to have encouraged individual managers to produce evidence of issues which are often neglected. For example, many of the individual informants openly reflected on their relationship to the technological discontinuity and to the other actors involved. Most essentially, however, the evidence collected has made it possible to conceptualise the boundaries between different managerial actors, their managerial objectives and more personal interests, which is essential to our understanding of how technological discontinuities are managed in firms.

The findings which will be examined in detail in the following chapters emphasise the importance of the gradually constructed and changing interests of actors and how they are fulfilled and reorganised in the initiation and implementation of technological discontinuities. On the one hand, this is due to the long-term effects that new technologies may have and on the other hand to the various opportunities that discontinuities of these kinds offer for individual and collective gain. For example, in addition to the fact that new production technology in the capital-intensive industries has a lifetime of several decades, the projects in which these technologies are built also form the basis for the professional reputation of the managers involved. Thus the long-term effects of investment have major consequences for the management of technological discontinuities in such contexts.

The contribution of the present study

The main argument of this book is that previous research has not adequately treated the issue of how divergent managerial actors become mobilised to overcome technological discontinuities within firms. The contribution of the book is thus twofold. First, it presents a theoretical longitudinal case analysis of how conflicting managerial actors become mobilised to co-operate in the initiation and implementation of a discontinuous technological change. The previous concepts and categories are used in the theoretical framework presented in the first chapters of the book. This framework is then further developed on the basis of empirical analysis. This analysis is innovative because in addition to depicting the various actors involved in the discontinuity, the study also recognises factors which encourage them to co-operate in its execution. This is done by explicating the relationships of the different managerial actors to each other and to the technological discontinuity. Moreover, the book elaborates on the mechanisms which make the mobilisation of organisational actors inherently fragile and susceptible to decline. This aspect has received little attention in the previous management literature which in most cases (e.g. Nadler *et al.* 1995) treats discontinuous change as a mechanically manageable matter. The conceptualisation of managerial actors presented in this book is thus more sophisticated than the analyses normally presented in the actor-focused management literature.

Second, the book also makes a contribution by introducing a contextual explanatory perspective to the management of technological discontinuities. The term contextual here means that we relate the long-term development of corporate activities to the happenings in its business environment. Thus in addition to approaching a theoretically relevant management problem this book also joins the emerging body of literature approaching firms and management as products of their institutional and national context (e.g. Dawson 1994; Clark 1995). Technological discontinuities are demonstrated to emerge in part from the resources inherent in the sectoral context of the firm. The longitudinal perspective adopted in the study makes it possible to identify how managerial actors utilise these material prerequisites to effect a discontinuous technological change. For example, the book shows how the solutions to universal management problems partly originate from the resources offered by the firms' local environment. This also means that the book aims to respond to both individualist and contextualist concerns in the study of the management of technological discontinuities.

The book is also related to the emerging literature on the nation-specific characteristics of management. The contribution of the book on this field is significant mainly because although there is an emerging body of European management literature (e.g. Hickson 1993), it has not covered the distinctive history of the Finnish corporate structure and management mobilisation. The forms of this mobilisation are most advanced in the paper industry, where Finnish firms are now international giants representing the technological leading edge in their field. In general, the specificities of the Finnish economy are of increasing interest, because Finland is now also a member of the European Community. As a consequence, Finnish firms are the key players in the European forest industry field. More importantly, however, the case of the Finnish paper industry is especially interesting in the sense that with limited resources, the Finnish economy has been able to generate technologically outstanding firms in paper-making. The nation-level perspective is therefore especially fruitful here because it is used to shed light on the processes through which nations with limited resources may create competitive industries (cf. Fynes and Ennis 1997).

The outline of the book

This book presents an inductively constructed conceptualisation of the management of technological discontinuities. This conceptualisation is based on longitudinal research which combines contextual perspectives with those of managerial actors in a particular technological discontinuity. The research assumes that management is a constellation of several actors each of whom holds distinctive, although gradually changing, managerial objectives and personal goals. Moreover, it is assumed that these actors become comprehensible only in the ongoing context in which they act. This is to say, managerial actors are groupings of individual managers based largely on the specificities of their current work environment. For example, subdivisions within management are dissimilar at times of corporate expansion and decline. This study therefore conceptualises managerial actors and their mobilisation to initiate and implement a specific technological discontinuity. From these starting points, the study aims to answer the questions how and on what conditions divergent managerial actors are mobilised to promote discontinuous technological change. The study therefore explores the ways the objectives and personal interests of managerial actors were combined with more substantial factors when Tampella effected a major technological discontinuity in its paper industry operations in the late 1980s.

The book is divided into seven chapters. The introductory chapter addresses the major themes which are developed later. It briefly sketches the relationship between subdivided management on the one hand and the requirements of the technological discontinuities on the other. It justifies the present study by tentatively positioning it in relation to the recent streams of research on this field. It also introduces the empirical setting of the study and briefly outlines its contents.

The two chapters following the introduction present the conceptual framework of the study. The second chapter reviews relevant streams of research within the previous organisations-oriented literature on technological change. The chapter begins with further reflections on the relationship between managerial action and technological change. It then identifies both the formal and informal factors which divide management into conflicting groupings. It proceeds to discuss the nature of technological discontinuities in capital-intensive industries and analyses the specific requirements of discontinuous technological change for managerial action. The chapter closes by arguing that the initiation and implementation of technological discontinuities is a major managerial challenge with a special need to mobilise managerial actors for intensive co-operation.

The third chapter goes on to examine how the mobilisation of managerial actors to initiate and implement technological discontinuities has been described in the previous literature. The chapter assembles a framework that synthesises the previous approaches on management mobilisation. It develops the argument that although the importance of mobilisation of managerial actors in the management of technological discontinuities has already been recognised, the previous literature has inadequately studied how, and why, actors with distinctive managerial objectives and personal interests become committed to the discontinuities embodied in major technological change projects. To justify this claim the chapter presents several approaches to the problem and shows how this study extends the existing body of knowledge on the mobilisation of managerial actors to initiate and implement technological discontinuities. For example, it discusses mechanisms related to the decline of mobilisation often omitted in the previous literature. It also discusses the effect of material resources on management mobilisation and makes a case for the claim that management should be conceptualised as a temporary constellation of conflicting actors and actors' coalitions with distinctive goals and forms of operation.

The fourth chapter examines the general prerequisites for the study of management of technological change and describes the research design and data on which the present study is drawn. It starts from describing the overall characteristics of the Finnish paper industry and positions Tampella within that industry. After presenting some reflections on the strengths and limitations of the adopted case study approach the chapter goes on to examine the field work procedures in more detail. This is important especially because research aiming to produce conceptualisation on management mobilisation poses significant methodological concerns. This chapter therefore pays special attention to how the researcher gained access to study the Finnish major corporation Tampella and the kind of evidence obtained in this work.

The fifth chapter of the book describes the necessary details of the research setting. That is, it analyses the features of the paper products, production technology and management of Tampella before, during and after the specific technological discontinuity in focus. The chapter provides readers with the essential aspects of Tampella's technological and managerial capabilities and the events of the actual change process in order to enable them to critically assess the more interpretative parts of the analysis presented in the following chapters. The chapter describes the development which led to the relative decline of the company and the situation of 'do or die' in which an ambitious technological change project was initiated and implemented. The closing of the chapter summarises the challenge of technological discontinuities to managerial action in this specific case.

The sixth chapter provides the actor perspective on the technological discontinuity in focus. It aims to clarify how the divergent managerial actors were mobilised to initiate and implement this technological discontinuity by analysing the constellations of managerial actors at each stage of the major technological change project. This is done mainly in the first section of the chapter, which recognises the formation of conflicting managerial groupings, their objectives and other characteristics and the conflicts between them. The chapter then proceeds to examine each actor in relation to the technological discontinuity at hand. This is done in the second section of the chapter which demonstrates the triggers of mobilisation which brought each managerial faction to support the discontinuity. After analysing how the managerial actors were mobilised the chapter goes on to examine how this mobilisation soon began to decline. The third section of the chapter therefore presents how a major technological change project, which at first permitted the actors to promote their managerial objectives and personal interests, soon became a potential threat to them. Finally, the chapter points out the more general implications of the presented analysis for our understanding of the intra-managerial triggers for and obstacles to the mobilisation of managerial actors. The chapter closes by summarising the major features of the mobilisation process in focus and by highlighting what aspects of the analysed changes deserve to be further examined in the remaining parts of the book.

The seventh chapter provides a contextual perspective to the management of technological discontinuities. By examining the background of the technological discontinuity in focus it aims to capture those conditions which allowed managerial actors to produce it. The explanatory framework includes analysis of the long-term traditions of developing technological capabilities through major technological change projects both in the paper industry in general and in the case of Tampella in particular. It also includes an analysis of the temporary changes in Tampella's corporate resources and business environment. The developed framework is then used in the further analysis of the case in focus and also of the Finnish paper industry sector in general. The evidence presented suggests that simultaneous changes in firm-, sector- and nation-specific characteristics may form motivating triggers for management to initiate and implement technological discontinuities. In short, the seventh chapter presents empirical evidence to justify a claim that the historically formed patterns of making technological change are recurrently materialised at times when both corporate resources and market position enable and encourage managerial actors to make changes.

The eighth chapter concludes the main themes of the book and reflects on the theoretical and practical implications of the presented analysis. Specific attention is paid to the inborn relationship between discontinuous change and the mobilisation of managerial actors. This means that technological discontinuities are only one example of a variety of changes that can be promoted by management mobilisation. The chapter also envisions some expected developments in the conceptualisations of these processes of mobilisation in general and in the study of the management of technological discontinuities in particular. The chapter closes with some reflections on the practical implications of the study.

2 Management and technological discontinuities

Technology as an outcome of managerial action

Technology and technological change have recurrently appeared in the main debates within organisation and management studies but the form in which the relationships between technology and management have been conceptualised has developed significantly in recent decades. For example, until the early 1970s organisations-oriented research on technological change was dominated by technological determinism. This means that technological change was considered an external determinant that compelled organisations to change (e.g. Woodward 1965; Thompson 1967; Perrow 1967). Technology was therefore not something to be intentionally modified but instead a contingency to which organisations had to adjust. For example, Galbraith (1974) found technological complexity to increase the amount of information processing. As a consequence, the co-ordinating activities of the organisation also had to be increased. In relation to the themes of this study this also means that technological change (both the technologies available to perform a given task and those actually selected for this purpose) produces alterations in organisational power structures and management styles.

In contrast, the actor-focused approaches have seriously questioned the deterministic view since the early 1970s. The emergence of actor-focused work on this field was partly related to the fact that despite the large body of literature attempting to verify the deterministic claims, evidence of the influence of technology on organisational structures and management remained contradictory (Barley 1986: 78; Burkhardt and Brass 1990). In other words, the technology determinists and contingency theorists could not demonstrate a mechanistic relationship between overall technological change and the organisational forms of business firms. The main contribution of actor-focused research on technological change has therefore been the argument that managerial actors have a major impact on the forms of technological change within firms (e.g. Child 1972; Pettigrew 1973; Webb and Dawson 1991). This change of focus has brought a large number of organisational determinants on technological change under study (e.g. Tushman and Rosenkopf 1992) and opened several new paths forward in this area.

In general, the actor approaches to technological change have been justified by arguing that technology is subject to managerial authority like any other strategically important factor in the organisation (Burgelman and Rosenbloom 1989; Weiss and Birnbaum 1989). While managers have power with respect to the choice of technology (whose modernity is an increasingly critical issue in the current corporate world) they must also respond to developments in it. Accordingly, studies of technological change must examine who is in power in order to understand how technologies in firms have developed until now and how they will develop in the future. Moreover, we can assume that changes in top management are likely to have an important influence on the forms of technological change adopted by corporations (Hambrick and Mason 1984; Bantel and Jackson 1989).

The studies conducted in this spirit have demonstrated that the amount and quality of managerial capacity, including for example the cognitive abilities and limitations of individual managers and their professional background and education, correspond to the forms of technological change in firms. For example, the nature of technological change initiated has been demonstrated to reflect the powerholders' professional (Daft 1978) and other characteristics. More concretely, the growing number of technical specialists seems to promote adoption of advanced technology (Ettlie *et al*. 1984). Other studies dealing with the effect of changes in management composition on technological change have demonstrated that the amount of demographic diversity within the top management teams affects the level of overall innovativeness of corporations (Wiersema and Bantel 1992; Goodstein *et al*. 1994).

In addition to being important in general, managerial actors also have been found critical at some special stages of technological change. For example, recent studies in this field of research have concluded that top management makes a difference especially in the introduction of the new technologies (Papadakis and Bourantas 1997). On the one hand, this is because top managers have a variety of reasons to be active advocates of technological change. For example, in addition to economic and technical objectives they may also use technology as a tool for organisational control (see e.g. Preece 1995: 60–64, for a review on this perspective). Consequently, top managers who constantly scan the corporate position see the acquisition of new technologies as a response to different kinds of external and internal pressures. On the other hand, top managers actively advocate new technologies because the highly visible and resource-intensive technological change projects depend mostly on their backing. For example, individual top managers have been found, when the corporate resources for innovation are limited, to be capable of bringing about such projects even without significant support from their subordinating organisation (Day 1994).

Whereas top managers in general have proved to be important promoters of new technology, those top managers who have only recently been appointed to their positions have proved to be especially active initiators of discontinuous technological change (Tushman and Romanelli 1985). Interestingly enough, this is so although newcomers in management, for example those recruited from outside a specific firm, often have rather limited opportunities for success as it takes time to gain sufficient understanding of the activities of each firm and to create the social liaisons needed for the formulation and implementation of technological change (Burgelman and Sayles 1986). Although they may have these handicaps, the newcomers may have experience which motivates them to initiate technological change in their new positions. For example, in their longitudinal study on Cadbury Limited, Child and Smith (1987: 576) showed how ideas for company transformation, including major technological changes, were brought by managers recently employed by Mars, one of Cadbury's primary competitors. Managerial turnover may therefore accelerate technological change, especially if the new recruits have previous experience in the same businesses. The newcomers may be familiar with the latest technology (Langley and Truax 1994) or they may be capable of applying existing capabilities in a new way (Kanter 1988).

Thus there is a large body of literature claiming that technological change in firms can be interpreted to reflect the characteristics and interests of the executive team (McGrath *et al.* 1992). However, the picture becomes more complicated when we incorporate another closely related body of literature assuming and demonstrating the usefulness and adequacy of regarding management of technology as essentially a political process. On a general level, previous studies in this stream have demonstrated that technological change is not only an independent process, but also a result of political negotiation (Noble 1984). This means that we assume that instead of one, there are several political interests involved in management and these firm-level political processes need to be conceptualised before a coherent understanding of technological change can be acquired (Barley 1990a). For example, major technological-change projects in companies not only involve deciding on whether a project is carried out or not. Instead, it can be argued that the internal power relations within a company largely decide which project is carried out: one project often supersedes others (Kanter 1988; Laurila 1997a).

To briefly summarise the ideas presented so far, it seems warranted to argue that management is the major source of technological change in firms. We have already noted that the impact of individual top managers is somewhat restricted by the fact that technological change is always incorporated through a political process. However, we do not contend that this is the only factor limiting the impact of top management on technological change. In contrast, it needs to be acknowledged that there are also other factors which significantly restrict and adjust the management effect on technological change in firms. Most importantly, there is a large body of literature – that has its roots in evolutionary economics – which has shown the inevitable bonds between managerial initiative and technological change. The seminal works on this field have demonstrated that although managerial actors are important sources of technological change, their impact is heavily conditioned by their past experiences and capabilities and fixed corporate routines (Dosi 1982; Nelson and Winter 1982).

Following this line of argument, organisation and management researchers have recently argued that management is basically unable to orchestrate innovative change (Pavitt 1991). The main reason for this is management's incapability to learn from previous change projects (Bessant and Buckingham 1993; Child *et al.* 1987). The same types of solutions are used repeatedly and technological change projects therefore may, instead of profoundly upgrading existing technologies, become more or less ritualistic activities. Some recent studies have shown that managers often repeat previous changes because the experiences gained in each project only marginally alter their habitual modes of operation (Tyre and Orlikowski 1994: 100).

In addition to being socially and cognitively restricted, managerial actors may also intentionally impede technological change and innovation in firms. This is because new technology often implies the threat of becoming incompetent or at least requires painful learning of new skills. Individual managers may rationalise this reluctance to innovate by referring to the similarities of products and production technologies. These similarities permit a variety of economies of scale and scope if the new resources are spent on extensions of the old businesses instead of on creating new ones. Applying the concepts generated in evolutionary economics, firms tend to follow specific technological trajectories in their technological development. This is because they tend to adopt capability enhancing instead of capability destroying innovations (Tushman and Anderson 1986; Dosi *et al.* 1988; Anderson and Tushman 1990). Organisational inertia is thus embodied in the technological trajectories of a specific firm or industry. These trajectories represent the logic through which management has learned to direct the technological development of the firm. In short, there are strong motivations for management to continue on the previously created tracks in the management of technology in general and in the initiation and implementation of technological change in particular.

Recent studies have specified the previous arguments concerning the management effect on technological change. Most importantly, they have explored the background of organisational inertia in more depth. These studies show that instead of acquiring completely new technologies, firms most often tend to develop their existing technological capabilities with the new technologies (Markides and Williamson 1996; Pennings *et al.* 1994). A similar finding, established from a somewhat different background, has been produced by Chen (1996), who noticed that firms tend to diversify in the direction of their existing activities. Using the ideas of classical innovation studies, these findings can be interpreted to indicate that managers do not adopt new technology that appears to lack compatibility with the old (Ettlie *et al.* 1984; Zaltman *et al.* 1973; Rogers 1962). This means that managers are sensitive to the inherent synergies between old and new activities – that is, how the technology acquired relates to the technologies already in use in the different corporate business activities – and tend to protect their organisations from too abrupt technological change. Moreover, it has been found that managers usually choose new technologies which will facilitate further innovations (Fennell 1984). To put it another way, good economic prospects are not sufficient justification for new business venture. Instead, only those projects with substantial promises for the long-term development of company activities have realistic chances of becoming accomplished.

Accordingly, it can be expected that management chooses new technologies that require capabilities substantially similar to the old. Although viable in the development of all business activities, this tendency can be expected to be especially strong within industries characterised by a large scale of production technologies or at least temporarily high levels of uncertainty. In this book we present evidence from the paper industry which in its ultimate forms is characterised by both of these features. It has been argued that in such contexts even clear environmental indicators of the need to enter new businesses do not necessarily outplay managerial commitment to the existing ones (Burgelman 1994; Stuart and Podolny 1996). This is not because it would not be worthwhile to enter into innovative businesses but because the existing technologies and technological capabilities encourage a search for new businesses mainly within the limits of the existing areas (Mitchell 1989; Chatterjee and Wernerfelt 1991). As a result, it seems warranted to argue that previous technologies enhance technological change in some business areas at the cost of others and thus significantly mediate corporate expansion into new businesses.

In summary, the above streams of literature propose that although management can be considered the major initiator of technological change in firms, its impact is seriously constrained by the existing routines and technologies. Thus on the basis of the research mentioned above we should expect that management initiatives were plagued by inherent inertia and managerial actors were rather consistent in their approach to managing technological change. It needs to be emphasised that this book somewhat questions such a claim. Instead, it argues that managers may have significant degrees of freedom when they initiate technological change. This is, first, because even technological change which intentionally builds on existing technologies can vary significantly. For example, it has recently been argued that capabilities-based technological expansion may take place in specific or parallel businesses (Boisot 1995) and the technology strategies adopted in such expansion may include at least niche, competence extension, relatedness development and 'thousand flowers' (McGrath *et al.* 1992). Second, it seems that the management effect on technological change is significantly conditioned by contextual determinants. For example, the longitudinal evidence on the Tampella Corporation presented in the latter parts of this volume clearly show that managerial actors with even minor previous experience may, in the context of increasing material resources and other supporting factors, initiate ambitious technological change that largely contradicts the company's previous traditions.

Finally, it is argued here that in order to understand why managers often initiate and implement discontinuous technological change, we need to be sensitive to at least two sets of factors not yet systematically covered here. First, instead of being a monolithic actor, management is divided into various subgroupings which makes the impact of management on technological change quite difficult to anticipate. Second, technological discontinuities require that subdivided management succeed in making quick decisions between several alternative approaches which, however, have long-term effects on both their corporations and their personal careers. This makes managerial judgement and choice extremely critical and emphasises managerial discretion instead of more deterministic forces. Both of these sets of factors are examined in detail in the following sections.

The nature of managerial subdivisions

This section examines factors that divide management into conflicting factions and thus make the impact of management on technological change highly unpredictable. The argument developed here is therefore that the management of a modern corporation is not a monolithic actor which aims to fulfil unified goals shared by all managers. Instead, it is divided by several formal, informal and situational factors which form numerous actors with distinctive interests and aspirations. This section does not, however, aim to produce an exhaustive list of possible sources of division within management. Instead, it only intends to illustrate the variety of these sources briefly. We begin from formal sources of management subdivision and then proceed to the informal ones.

Formal hierarchy as a source for management subdivision

The most important formal source of management subdivision is the vertical and horizontal division of managerial labour. On the vertical side, we have the differences in the amount of power and the content and purpose of action between managers representing different management levels. For example, managers at the lower levels of hierarchy are concerned with operational procedures whereas their supervisors and their supervisors' supervisors are responsible for business and strategic management and in the end for the creation and maintenance of the entire corporation (Teulings 1986). Although the importance of the formal management levels for the managerial actors themselves is difficult to assess from the outside, the visibility of these vertical divisions have made them a feasible object of empirical study. For example, recent studies on the strategy domain have problematised the relationships between corporate boards and CEOs (Boeker 1992; Cannella and Monroe 1997) and between top management and the middle managers (Howell and Higgins 1990; Fulop 1991) in initiating corporate change. These studies have shown that although middle managers have a limited amount of formal power, they have an important role as advocates of corporate entrepreneurship and innovation.

On the horizontal dimension of formal hierarchy, we have the various functional departments with their divergent objectives and focus of attention (Lawrence and Lorsch 1967; Armstrong 1986). An important fact increasing the importance of these lateral hierarchical divisions is that each functional department – like production, personnel or sales – has specific variables it seeks to optimise. The boundaries between different functional departments are also enhanced by the fact that each of them tends to be manned by individuals representing specific professions. These professions have their own cognitive models and linguistic codes created through common education and other profession-specific forms (see Van Maanen and Schein 1979; Van Maanen and Barley 1984; Torstendahl 1990). Managers in a specific field or function often have the power to recruit and dismiss new employees in their field. Managers with specific education may benefit in this way from keeping a specific function under the control of those with similar education. For example, engineers may dominate not only production and production development functions, but also the finance and sales departments of technology-driven firms. This is not, however, to say that managers sharing a particular occupational background necessarily agree. Instead, intra-professional groupings may be based for example on the universal differences between cosmopolitan and local bases of loyalty (Gouldner 1957). Moreover, the same managerial community may include several factions supporting divergent operational traditions (Strauss *et al.* 1963; Bierly and Spender 1995; Parker 1995).

Managerial subdivisions based on formal hierarchy are thus especially vital because the differences between the formal responsibilities of managerial actors are coupled with differences in their career paths, methods of operation and sources of power. The picture of management subdivisions becomes even more complex in the context of the multidivisional corporation. The creation of business units leads to a differentiation between line and staff management (Dalton 1950). Because of the differences between the functional expertise of staff managers and the general skills of line managers, these groups often remain rather isolated from each other (Crozier 1964). For example, in current industrial corporations it seems that whereas managers who have long worked in line positions are often appointed to staff positions at later stages of their careers, the opposite procedure is much less common.

There are also other ways in which multidivisional corporate structures enhance the above mentioned horizontal divisions within the managerial hierarchy. This is especially because these divisions follow the lines of the different ranges of business which compete for scarce corporate resources. In those multidivisional companies where business divisions differ a lot, the specificities of industry, such as technology, are one additional source of social and cognitive boundaries within management (Pennings and Gresov 1986; Gordon 1991; Sackmann 1992). Because it is inevitable that all businesses cannot be the core businesses at least in the long term, the managers representing each business division have to compete. This kind of intra-managerial activity has been called politicking (see e.g. Mintzberg 1985: 134–139, for a review) and leads for example to empire building and suboptimisation within the corporations.

Informal sources of subdivision

In addition to being formal action towards clearly-defined objectives, management also includes informal social action (Dalton 1959). This informal action is relevant here because although these divisions are difficult to observe from outside the company, they can be expected to be even more vital than the divisions produced by formal hierarchical divisions. This is especially so when the formal systems of authority are weak or unclear and when the conditions otherwise encourage such informal alliances within the corporation.

In general, the informal divisions within management are strong because they connote with the emotions, personal feelings and values of individual managers simultaneously associating them with some colleagues and disassociating them from others. To put it more concretely, when talking about informal action in management we refer especially to formally undefined relationships between actors which are most often based on commonalities of experience between two or more managers. Social cliques which have been defined as informal associations of two or more persons to realise some end (Dalton 1959: 53) are a classic example of such informal social relationships. Groupings of these kinds are based on emotional bonds within a group of managers and they may emerge in vertical (between superiors and subordinates) or horizontal (between formal equals in different departments) forms.

The importance of the informal intra-managerial boundaries for the relationship between management and technological change is based on the fact that they may create unofficial power centres within the formal management structure. Those interests which could be identified from the formal divisions are therefore only part of the political processes conditioning management influence on technological change. In the framework proposed by Dalton (1959), cliques function as reciprocal (though not necessarily equal) relations in which services are exchanged. Unless we know about such relations, many aspects of technological change remain obscured. This is not to say that such informal relations would always be unobservable for an external witness. For example, we can expect that correlated moves of several managers from one firm to another may often reflect the existence of an informal political coalition. In such a coalition a group of managers helps each other in the form of appointments to key positions in a specific managerial hierarchy.

We have thus argued that informal social boundaries are vital and may also have unexpected consequences for technological change in firms. It needs to be acknowledged that although informal managerial groupings and networks may have malfunctional effects, they are also an important tool for managers in their work (see e.g. Kotter 1982). This is because such informal liaisons may connect managers within a single management hierarchy or between separate firms. It seems evident that socio-emotional ties of these kinds may permit transfer of knowledge between actors and co-operation which would be impossible on the basis of formal relationships alone. For example, through informal coalition forming the competent but powerless actors at the lower levels of the managerial hierarchy may exceed their formal power and make their knowledge available for the use of the whole corporation. According to a prescriptive model of this process (Quinn 1980), managers at the lower levels of the management structure may incrementally rise to a major position through their up-to-date knowledge, coalition forming, and experimental success in business operations (see also Burgelman and Sayles 1986).

Another perspective to the informal management subdivisions has been provided by organisational culture researchers. From their perspective, monolithic management would require a patterned system of values and meanings, making individuals susceptible to specific ways of managing (cf. Golden, K. 1992: 5). Because individual managers can (either openly or un-openly) oppose such patterns, managerial subcultures which in some way deviate from this general pattern emerge. For example, because there are almost always competing technologies for an individual task, it can be expected that each of them will have their own supporters within management. In the organisational culture framework such phenomena indicate that although managerial hierarchies often carry a consistent cultural image, we can find a variety of practices and meanings within them (Meek 1988; Jermier *et al.* 1991; Trice and Beyer 1993). Accordingly, disagreements on the content and timing of technological change are only one reflection of cultural diversity within organisations. We also should expect that even a group of managers that looks homogeneous may contain significant contradictions. This is because not all actors necessarily pursue similar objectives or interpret joint acts (e.g. supporting the introduction of a specific new technology) in the same manner (Young 1989).

In summary, managers are divided in many senses and for many reasons. Management can thus be considered to consist of conflicting factions which treat concrete managerial problems differently (Harris 1994) and which find different ways to promote the general managerial objectives they all accept. As a consequence, decision making on the content and timing of technological change is a process to which all of these actors with their specific knowledge and power bases can contribute. However, before this view can be further developed we need to examine the other of the two above mentioned reasons for the unanticipative nature of management impact on technological change: that is, the specific requirements of technological discontinuities for managerial action. In fact, the conceptualisations presented above are generic in the sense that they do not make a distinction between phases of technological discontinuities and incremental development of technology (e.g. Tushman and Romanelli 1985). Moreover, they deal with management only in a universalistic manner which is not sensitive to the features and traditions of specific firms, industries or national contexts. The next section therefore examines the challenge technological discontinuities pose for managerial action in general and in the context of the contemporary paper industry in particular.

Technological discontinuities as a managerial challenge

The nature of technological discontinuities in capital-intensive industries

We have, above, defined technological discontinuities as situations in which the product or production technology of the firm is altered in a way that makes the previous capabilities obsolete and requires the development of new ones on the different organisational levels. Technological discontinuities thus imply that either within a firm, or outside it, some kind of a technological breakthrough takes place and is then utilised in the form of new technologies bringing about new businesses. Previous studies on this field have argued that such breakthroughs and revolutionary innovations are rare (e.g. Tushman and Anderson 1986; Anderson and Tushman 1990). Our experience of the paper industry confirms this argument. Radical innovations in production and especially in product technology do not happen often. However, this is the case only as far as the whole industry is concerned. In contrast, when looking at individual firms and their production units we meet technological discontinuities much more frequently. This is especially because individual firms usually have to manage discontinuities between the old and the new technologies whenever they replace or rebuild their production or product technologies.

Despite the fact that only some of the individual managers ever have to head a major technological change project, they are an essential and increasingly typical feature of current corporations. For example, in large paper industry companies with several production facilities, there are usually at least some projects where old technologies are being replaced with new technologies of significantly more advanced properties. Individual paper industry firms usually replace their production technologies (i.e. paper machines) only after their twenty or thirty year life-cycle has passed. In recent decades, however, various kinds of rebuilds and renovations within this period have become common. As a consequence, before becoming totally exhausted the machinery may already have been rebuilt two or three times. Consequently, various characteristics of the technology (e.g. speed, paper grade, the characteristics of the refinement process) have been significantly altered. This is partly because even minor changes in the quality of the produced paper usually require major modifications in the production technologies. Moreover, although the paper grade remains the same the new technology (because of the continuing technological development) is quite different from the old. For example, newsprint as a product has had roughly similar properties for decades, although the technologies used for its production have more than tripled in scale. Thus the change from the old to the new technology is often fundamental.

As in any other capital-intensive industry, the development of technologies in the paper industry is highly path-dependent. Although managers have opportunities to choose between different paper products and different production technologies to produce the same products, they nevertheless usually (at least as far as brownfield investments are concerned) have to fit the new technology along a complex and incrementally developed chain of technical operations. In other words, the vertical integration of the production process in the paper industry (see Davis *et al.* 1992; Globerman and Schwindt 1986; D'Aveni and Ilinitch 1992) requires that the new elements of production technologies must be compatible with the old unless the production facility is new or the entire line of production is to be replaced. For example, the specificities of pulp production define the portfolio of possible new paper grades in a paper producing company. Moreover, paper production consists of several distinctive capabilities (e.g. paper bleaching, calendering and coating) which are built on each other and through which the paper becomes more and more refined.[1]

Despite their salience, these path-dependencies do not, however, diminish the importance of technological discontinuities. This is merely because no new technologies exist that would be directly compatible with the old. For example, it is always problematic to replace an old paper machine with a new one in the sense that the rest of the facility often uses technology produced decades earlier. Consequently, the replacement of one element in a continuous chain of production process always poses a technological discontinuity from the management perspective.

In summary, technological change in paper industry firms include unavoidable discontinuities in addition to continuous incremental adjustments. These discontinuities are an extremely important management issue because in capital-intensive industries with largely standardised products, managers have little influence on short-term corporate performance. This is due for example to unpredictable fluctuation in factor prices, imbalances in supply and demand leading to the price volatility of the products, and foreign exchange fluctuation. Most importantly for the themes of this book, however, these particulars bring technological change to the focus of managerial action both in the form of minor improvements and major technological change projects.

Technological discontinuities as a managerial task

The initiation and implementation of technological change in firms is always a challenging mission for management. This is because there is no one single relationship between the overall technological development and technological change taking place in firms. Instead, management has to consider several alternative technologies and technological solutions whenever it decides to introduce new technology. In other words, managers always have several alternative approaches and only limited opportunities to anticipate the outcomes of each approach. We argue that these ambiguities become especially critical in the management of technological discontinuities.

There are several reasons why technological discontinuities place specific demands on managerial actors. First, this is related to the high risks involved in such discontinuities. For example, the capital needed to build or rebuild a single paper machine may exceed the yearly turnover of several such units. The paper industry companies must make greater and greater investments as the typical size of paper machinery constantly escalates. The risks involved also increase because the building and start-up of a new or a rebuilt production facility include several uncertainties. For example, the quality of the paper produced and the efficiency of the new machinery depends on the abilities and co-operation of both the technology supplier and the personnel of the paper mill. Thus it is not possible to guarantee the technological success of a new facility beforehand.

Second, technological discontinuities pose an ambiguous task for management because of the difficulties in selecting and defining the type and properties of the new technology. For example, the manufacturing of specific paper grades requires specific production technology, which means that each time a paper machine is rebuilt or replaced, management must decide in which paper grade the investment will be made. The paper grade chosen and its market situation affects the start-up of the new machine and sets limits on its efficiency and profitability. Another critical issue in technological discontinuities in the paper industry is connected to their timing. The fluctuation of demand and supply in general and in specific products, and the long time needed from the investment decision to the actual start-up makes it impossible to define a single optimal moment for a major technological change project. Moreover, the feasibility studies made for an individual investment often go wrong because paper industry companies throughout the world tend to extend their production capacities at the same time.

The specific characteristics of technological discontinuities are relevant here especially because their inherent ambiguities pose a major problem especially for the subdivided management. We believe that this problem is significant for both researchers and practitioners. This is especially because formulating and implementing a major technological change project would be difficult even for a unified group of actors. The difficulties are crucial when the actor is in fact divided in many ways. The previous section of this chapter demonstrated that management can and should be considered a composition of conflicting groups of managers who hold divergent managerial objectives and personal interests and preferences concerning their realisation. Despite these various sources of conflict, managers have to repeatedly reach agreement on which old technologies are replaced or rebuilt and which new technologies are adopted. Moreover, they have to agree on the parts of the company where the changes take place, how the various necessary capabilities are acquired, how much time is allocated for the changes and how the old and the new technologies can be made sufficiently compatible.

In short, it can be expected that such a composition of divergent actors will not, without specific facilitating mechanisms, reach an agreement on the content and form of technological discontinuities. The previous management literature has acknowledged that managing technological discontinuities necessitates at least temporary change alliances and coalitions (Kanter 1982) and commitment to their specific objectives by management (Burgelman and Sayles 1986). In general terms, this is because intensive technological change increases the internal interdependencies within managerial hierarchy (Burns and Stalker 1961; Lawrence and Lorsch 1967). Using the words of Hambrick and Siegel (1997: 26) 'technological intensiveness imposes a considerable requirement for collaboration among senior executives'. In the case of a paper industry company this means for example that there is a need for co-operation between mill managers who possess business-specific knowledge and who are usually responsible for formulating the projects and the top executives who are the key power-holders in the organisation and who are formally responsible for the investment decision.[2]

Additionally, the need for the mobilisation of managerial actors to initiate and implement technological discontinuities is further increased because the existing subdivisions of management, in fact, tend to be reinforced in such conditions (Pettigrew 1973; Bloor and Dawson 1994). This means that the diverse ambiguities connected to the technological discontinuities make the different managerial views of their content and timing especially vital. Moreover, the fact that the group of managers involved in discontinuous technological change projects is large, increases the probability of conflict (O'Reilly *et al.* 1989). For example, merely formulating and deciding on a major change project in a paper industry company involves co-operation throughout the management structure. The probability of conflict is also high because the way technological discontinuities are formulated largely determines which become or remain the core businesses and which are relegated to peripheral status in the corporation.

Later in this book we will empirically demonstrate how a specific technological discontinuity emerges as a consequence of a mobilisation process creating temporary shared commitment among divergent managerial actors. On the basis of the details mentioned above, we argue here that the specific problems of facing and producing these kinds of discontinuities makes the paper industry an interesting context for the study of management mobilisation. We even go on to argue that the Finnish paper industry is particularly interesting in this sense. We already noted that the distant location from the main international markets has forced the Finnish paper industry firms to use technological change projects to become major international corporations in their field.[3] Consequently, the Finnish paper industry has become a high-velocity environment (Eisenhardt and Bourgeois 1988) with specific requirements on managerial action.[4] Most importantly, the systematic use of technological discontinuities as a tool to increase and sustain competitiveness has made mobilisation of management a key issue in the industry. We therefore believe that an in-depth analysis on how a Finnish paper industry firm has managed this issue is of high relevance for further theorising. However, before proceeding to a full empirical examination of these issues, the next section will review how previous literature describes mobilisation for discontinuities by management as a subdivided constellation of actors.

3 Mobilisation of managerial actors for discontinuous change

Above we described the specific characteristics of technological discontinuities and concluded that mobilisation, that is the development of mutual commitment between divergent actors, is a critical issue in managing such discontinuities. In this section we examine how this process has been described in previous management and organisation literature. However, before moving on to a review of the literature, let us turn to research which could be used to question or even deny the existence of the entire problem in focus here.

Several studies argue that managers need not be mobilised to initiate and implement discontinuous technological change. In other words, it could be assumed that although conflicting managerial actors exist, specific mechanisms for overcoming their conflicts are not needed. This is, initially, because managerial actors not only work to fulfil their managerial objectives but they also work for their living. They are therefore motivated to find ways to combine even contrasting objectives into collective action (Weick 1979). Moreover, whereas we argued above that the operational pressures of technological discontinuities require managerial co-operation, it could also be assumed that such operational pressures virtually compel actors to co-operate (Young 1989: 203). Consequently, no specific mobilisation mechanism would be needed. Likewise, managing a technological discontinuity by no means requires collectively shared objectives and interpretations within the managerial hierarchy (cf. Grieco 1988). This is partly because the shared means and conventions of communication also decrease the amount of consensus needed for organisational actions (Cohen *et al.* 1972; Donnellon *et al.* 1986).

The literature mentioned above suggests that individual managers and managerial actors may be inclined towards mutual co-operation. However, the problem with this kind of approach is that it cannot explain why managerial actors take part in action which requires commitment, significant personal sacrifices and co-operation with actors representing dissimilar values and ideologies. For example, the managers involved in a major technological change project must accept the risk of project failure and its consequences for their future careers and work with colleagues of a completely different background.

We therefore argue that without conceptualisation of the mobilisation phenomenon we cannot illuminate how conflicting managers, for example in a paper industry company, reach agreement over the timing and design of a large-scale technological change project. Accordingly, we consider the problem of mobilisation highly relevant and review different kinds of approaches to that problem in the following sections. We begin from the impact of individual top managers and their personal characteristics and then proceed to the conceptualisations of management coalitions. In the previous chapter we already noted that managerial actors are an important source of technological change in corporations. In the following section we will examine this aspect from a somewhat different point of view and show how individuals and groups of managers can mobilise others to produce discontinuous technological change.

Management skills and personal charisma

There is much evidence to justify the argument that the personal skills and characteristics of top managers influence various organisational outcomes (e.g. Hambrick and Mason 1984; Finkelstein and Hambrick 1996; Thomas *et al.* 1993). According to the general line of thinking in this body of research management, knowledge and expertise are the critical determinants of the variety in corporate strategic choices (Child 1972; Eisenhardt and Schoonhoven 1990; Castanias and Helfat 1991). Usually the studies within this stream have therefore paid attention either to managers' personality characteristics or to their expertise created through education and experience from the specific industries of this firm. The studies which have emphasised managers' personality characteristics more than their expertise have demonstrated that the top managers' need for achievement (Miller and Droge 1986), cognitive styles (Nutt 1986) and locus of control (Miller and Toulouse 1986) have an impact on organisational change.[1] For example, we may expect that managers with a high need for achievement most probably initiate ambitious technological discontinuities. The problem with this stream of research for the purposes of the analysis advocated in this book, however, is that although it has demonstrated that managerial characteristics do make a difference, it does not clarify how change is actually brought about and especially how the characteristics of top managers persuade conflicting managerial actors to co-operate.

In contrast, another stream of literature (sometimes called transformational leadership (see e.g. Cannella and Monroe 1997)) has more systematically examined the processual features of management-initiated change. In general, this body of work argues that the ability of top managers to mobilise others is conditioned by the relationship between them as leaders and their subordinates as followers (Yukl and Van Fleet 1992). This means that top managers have power to mobilise but their power is always conditioned by the fact that their subordinates continuously assess their skills and capabilities (Finkelstein 1992). This also means that the emergence of charismatic leadership, that is managers with an almost magic ability to make others committed and enthusiastic implementers of their initiatives, is always related to the original preferences of those being led.

Studies extending the early work by Weber (1947) on charismatic leadership have found that to mobilise others to support a specific change project, managers need to anticipate and utilise their subordinates' wishes and aspirations and to include them in some form in the project (Trice and Beyer 1986; Bass 1985). Thus it is assumed that through their persuasive action, key individuals may act as liaisons between conflicting managerial groupings through ongoing negotiation and by providing adequate rewards and messages for each faction (Trice and Beyer 1991: 163–164). Individual managers can therefore mobilise others, but only by taking their particular interests and aspirations into account. Moreover, the success of individual managers in their effort to bring about change appears to depend on their ability to create a positive interpretation of future events (Westley and Mintzberg 1989). Thus, each subcultural grouping needs to be contacted with messages 'framed' in a way that secures its commitment to, or at least participation in, the co-operation (Snow *et al.* 1986). This is possible because the features of a forthcoming technological discontinuity are always at least somewhat ambiguous. An inspiring vision of the implications of such changes may therefore partly promote its own realisation (Gagliardi 1986; Field 1989).

To summarise the argument so far, on the basis of the above mentioned literature, it can be argued that both the managers' actual skills and capabilities and their personal charisma affect their power to enforce co-operation within the managerial hierarchy. While top managers need their technical skills in formulating a technological change project, they are also highly dependent on their subordinates' aspirations and their assessed stature (Pettigrew 1972) when they seek to have their ideas applied. In other words, top managers are relatively powerful in initiating technological discontinuities, but their power is significantly more limited in the implementation of these discontinuities.

An entrepreneurial approach of this kind can be criticised for its inherent assumption that the capabilities of individual managers are sufficient to reconcile organisational conflicts. This assumption is problematic because these processes of managerial persuasion are time-consuming, individual managers have varying amounts of skill (Hosking 1991) and it is questionable whether charismatic individuals who can sustain their charisma will be found (Trice and Beyer 1986). For example, we believe that it is virtually impossible for the individual managers heading major technological change projects to be completely sovereign in their jobs and thus capable of maintaining a charismatic relationship with their subordinates.

Moreover, because charisma requires the formulation of ambiguous, risky or even contradictory objectives (Hater and Bass 1988; Bird 1989; Hill and Levenhagen 1995), such projects may be easily criticised and they may often fail to meet their objectives. This increases the probability that some actors oppose and act against major technological change projects either implicitly or explicitly. There is even evidence demonstrating how the efforts to advocate radical change projects produce new conceptual tools which organisational actors can use to legitimate oppositional acts (Dahler-Larsen 1997). This means that the rhetoric originally used to create enthusiasm and commitment to technological change may turn against their original aims. Charismatic leadership therefore should be considered a transient phenomenon, conditioned more by the particular circumstances than the personal characteristics of individual managers (Westley and Mintzberg 1989). Consequently it seems adequate to suppose that acts of individual managers are only one highly restricted mechanism of organisational mobilisation.

Political coalitions and combinations of managerial actors

The perspectives emphasising the impact of individual managers on technological change can be extended in different ways. The first opportunity is to replace individual managers with top management teams as the unit of analysis. An example of this kind of alteration is the literature on the composition of top management teams and its influence on organisational change. The studies in this stream have demonstrated that the educational and functional background (Wiersema and Bantel 1992), tenure (Finkelstein and Hambrick 1990) and diversity (Bantel and Jackson 1989) of top management teams greatly affect the emergence of managerial initiative for change. In fact, the findings are analogous to those presented in the previous section with the difference that instead of individual managers and their characteristics, we now speak about groups of managers.

Although the replacement of individual managers with management teams creates a more complex and diversified picture of the relationship between management and technological change, these studies are, however, rather limited as far as the mobilisation of conflicting managerial actors is concerned. This is especially because they conceptualise management mainly as a collection of individuals. Examples of types of coalitions between individuals within formal management structure are illustrated in Figure 3.1. The main problem with attention to groups of individuals is that it neglects the more collective aspects of managerial action. Such an approach is also relatively insensitive to the various formal and informal groupings within managerial hierarchies. In this sense a more fruitful perspective is offered by the political coalition approach (March 1962; Zald 1970). This approach assumes that the management structure of a firm is a changing composition of actors with both specific interests and sources of power. That is, all actors are influential although the amounts of power and the sources of this power vary. As a consequence, the objectives of a corporation are not considered to be given but instead negotiated.

In general, political perspectives on management assume that managerial hierarchy comprises several actors who are capable of promoting their own interests. Studies applying this tradition to the management of technological change have demonstrated that because managerial actors do not necessarily agree, the technology introduced is the one whose supporters succeed in marshalling the critical power resources (Pettigrew 1973). This is not to say that top managers are always the most influential actors. This is because, although top

Vertical

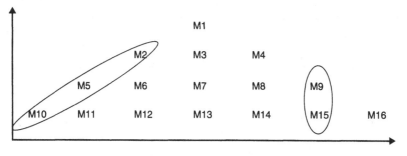

Figure 3.1 Examples of types of coalitions between individual managers within a formal management structure★

Note: ★M1 to M16 are used to indicate individual managers on different levels of the managerial hierarchy

managers have the utmost formal power, their subordinates often possess the latest knowledge on the field. However, from a political coalition perspective the mobilisation of conflicting actors to support a major technological change project requires that it offers benefits to all actors involved. In this spirit, discontinuous technological change may take place because in itself it may entail various positive outcomes for those who participate in its formulation and implementation. For example, besides the usual official objectives for new technology, such as increasing profitability or gaining competitiveness, an emerging business exploiting new technology permits 'scientists' to increase their knowledge and 'business people' to expand their businesses (Burgelman and Sayles 1986).

The political coalition approach thus predicts that whenever a coalition holding a critical amount of power and expertise is created, technological change can emerge. Various problems of implementation then arise, however, because the concrete change projects require the co-operation of virtually everyone. For example, although the lower-level managers or even shop floor workers involved may significantly hamper the flow of a technological change project, they could be considered minor players for the whole endeavour. It can therefore be argued that no modification of the internal balance of power will automatically allow mobilisation of the necessary capabilities and dispersion of knowledge throughout the hierarchy. This is not only because technological change involves creation of a vision of new technology, but also because of the concrete knowledge and capabilities that must be developed before the new technology can be adopted (Fennell 1984; Pettigrew and Whipp 1991).

Concrete problems like the difficulties in recruiting new technically competent personnel (e.g. Lee and Allen 1982) are a good example of the limitations of the political approaches to the management of technological discontinuities.

The interests of the conflicting actors may coincide, leading to the initiation of a technological discontinuity although the same actors lack the capabilities necessary to implement the change. The most significant limitation of the political approaches, however, is that they do not tell us how the necessary capabilities are created for those who are not committed to the technologies chosen. For example, it has been noted that even among the managers involved there are usually some 'foot-draggers' who negatively criticise the changes at hand (Kanter 1982). Although this may be due to the exclusion of these managers from the formulation phase of the discontinuity, such phenomena may threaten successful accomplishment. Consequently, changes in the internal balance of power cannot adequately explain how conflicting managerial actors become mobilised to support a specific change project. To do so, such mechanistic perspectives must be coupled with processual approaches based on more sophisticated assumptions of the social characteristics of management. Examples of these are outlined in the following section.

Processual approaches on management mobilisation

Previous literature exploring the relationship between managerial action and technological change, and which can be assigned the processual label, include the contributions of several researchers. These individual contributions have a few features in common. First, they assume that technological change has both instrumental and emotionally-laden meanings for the actors involved. This means that in addition to the political aspects of management, this approach also takes into account a variety of other forces behind the conduct of individual managers and their groupings. For example, whereas management has traditionally been conceptualised mainly through its general objectives and functions (Willmott 1987), the studies in this stream also assume that the specific features of managerial occupation (cf. Van Maanen and Barley 1984) can act as the basis for the motivation of the individual managers.

Second, the contributions in this stream assume that several coalitions are needed to bring about discontinuous technological change. Thus to both initiate and implement a major technological change, project coalitions between different actors in different roles are needed in the different stages of change. On the one hand this is because the sources of motivation for different managers vary. In a recent study Sturges (1997) identified the following overall career orientations of managers: climber, expert, influencer and self-realiser. Although we believe that managers with all these orientations can find ways to fulfil their interests in major technological change projects, we also believe that being involved in such projects would have different meanings for actors representing each of these types. We could expect that climbers would consider such a project mainly as a tool for their personal career advancement in the formal hierarchy. In contrast, managers representing the expert type would consider a major technological change project mainly as an opportunity to feel competent on the job and also to obtain recognition for this from others. Influencers would evaluate a change project mainly on the basis of its contribution to their more general managerial objectives and self-realisers would mainly pay attention to the personal fulfilment that involvement in such projects facilitates.

On the other hand, different kinds of coalitions are needed at different stages of a technological discontinuity because the requirements for managerial action differ at each stage of change. Apart from the fact that managers have personal preferences and capabilities which influence what kinds of activities they want to be involved in, there are different kinds of opportunities for these activities in different phases of a technological change project. For example, managers willing to be involved in the concrete building and start-up of a paper machine may find it frustrating that many times that phase only lasts a few months in a two-year project. Similarly, managers wanting to do the visioning for major technological change projects may soon notice that projects principally planned in a few weeks, may take several years to implement. However, in order to provide a schematic view of the emergence of technological discontinuities as a managerial process, we will describe below the different stages and sequences of a technological discontinuity and the role of different kinds of managerial actors in each stage.

The process of mobilisation of managerial actors to discontinuous technological change can be divided into four stages. The first stage is the initiation and formulation phase during which different kinds of actors need to combine their skills in order to provide a competitive change design. This must happen although these groups are at the same time competing within the same managerial hierarchy. For example, an emerging business exploiting new technology necessitates co-operation among scientists who approach the project analytically and also as an intellectual challenge, and business people who have a more straightforward attitude and who also sell the project to the rest of the company by utilising the knowledge created by the scientists (Burgelman and Sayles 1986). The enthusiasm of the first innovators is therefore based on both the anticipated benefits related to the major project and the satisfaction emerging from intensive work on such a challenging venture.

The enthusiasm created among the key initiators is also critical for the emerging project in the second stage, which can be called resource acquisition. More concretely, when moving into implementation the necessary material and personnel resources for the emerging project must be secured. At this stage new combinations of managerial skills become essential. This is because to bring about new businesses managerial actors must be able to align the particulars of the production technology with the market requirements and the structural context of the corporation (Bower 1970; Burgelman 1983). Because the ways managers formulate and present the new project affect its chances to obtain the necessary resources, often actors who do not know each other closely need to co-operate. Major technological change projects frequently require negotiation between management and trade unions or other local authorities. In such situations staff departments that normally are in a marginal position may become crucial for the completion of a technological change project. Moreover, the securing of resources necessitates demonstration of the potential of the project through managerial commitment expressed and reinforced through setting and reaching ambitious objectives. The technical properties of the new paper machines are normally announced publicly long before the machine is started up. This kind of strategic forcing (Burgelman and Sayles 1986) may further the social momentum needed to surpass the obstacles to the implementation of a major technological change project.

When moving to the third stage, which can be called implementation, the action around the change project has already become intensive and visible throughout the corporation. In successful cases this means that high internal and low external mobility is coupled with open communication and co-operation between actors from different backgrounds (Kanter 1983). In this study we will show how engineers with extensive previous experience of the current technology, at least for a short while, co-operated with engineers with significantly less previous experience in order to implement a major techno-logical change project. According to Walton (1980), the basis for such 'high commitment work systems' are meaningful work and a humane working envi-ronment in which it is possible to encourage spontaneous expressive commitment. Such commitment and enthusiasm is indicated by emerging revolutionary momentum (Miller and Friesen 1980) and by by-passing the formal decision processes through various forms of intra-managerial negotia-tion and persuasion (see Child and Francis 1977; Child and Lu 1996).

However, such commitment and enthusiasm can also easily be ruined. One key problem of producing discontinuous change in a large corporation through major development projects, is related to the fact that on the one hand the support of top management is needed to secure the necessary material resources, while on the other hand the independence of the project work must also be secured (Burgelman and Sayles 1986). In other words, being in the 'spot-light' of top management may be a hazard for successful project implementation. Especially those projects which have been granted major resources are susceptible to top management interference, which may lead to loss of momentum. In this book we will show how a negative reaction by the top management to shortcomings in the early stages of a major technological change project severely weakened the commitment of the original project champions. In addition, the momentum involved in a technological disconti-nuity is endangered by the fact that all actors are not similarly committed to the project. The managers trying to secure both the necessary project independence and the necessary resources have been called 'group leaders' (Burgelman and Sayles 1986). Thus successful implementation of a technological discontinuity necessitates the continuing contribution of these committed entrepreneurs and their corporate level sponsors (Kanter 1983).

The fourth stage of the mobilisation process can be called outcome. In successful cases it brings the actors those benefits they anticipated in the early stages of the project. Being involved in a successful project creates a positive reputation and leads to appointments to new managerial positions. Successful venturing provides opportunities for individual managers to improve their own position and that of their businesses in large corporations (Burgelman and Sayles 1986). Even more importantly, however, successful projects also create personal feelings of competence and pride and identity for the managers involved as experts in their fields (cf. Sturges 1997). Thus the motivation of individuals and groups of managers to work for a specific project is partly based on personal achievement. For example, the managerial rank connected to a group of managers for the rest of their working careers may be created in a single successful project.

To summarise, the most common solution to the problem of mobilisation can be called the two-centre theory (Day 1994; Witte 1977). In short, it argues that major technological change projects emerge when competitive ideas rising from the bottom are aligned with power from the top. In other words, competent business-level managers, called for example champions, entrepreneurs and souls-of-fire (Maidique 1980; Quinn 1979; Stjernberg and Philips 1993) convert operation-level knowledge into seductive action plans which are then sponsored by corporate-level managers (Burgelman 1988; Dean 1987). Thus the various tangible and intangible benefits involved in technological discontinuities temporarily unite conflicting managerial actors. As a consequence, persuasive and compelling action by top managers is not the key trigger for the mobilisation of managerial actors. Instead, coalition-forming between actors possessing different kinds of capabilities and power positions within the managerial hierarchy should be understood as a consequence of a process in which both rationalistic calculations and less rational emotional aspects are involved.

Despite its significant strengths, the previous literature on management mobilisation should be examined critically. One of the contributions of the above mentioned studies is the location of the basis of intra-managerial co-operation in the specific content of managerial work. However, it can be argued that the main body of the literature emphasising the mobilisation of managerial actors as a key prerequisite for technological change (e.g. Pettigrew 1973; Kanter 1983) has not made an explicit distinction between the effect of managerial objectives and the more personal interests and aspirations of the actors in the mobilisation process. Though the impact of the personal interests of managers has been recognised (e.g. Burgelman and Sayles 1986: 66), the way they are entwined and connected to the actors' managerial objectives has not been made explicit.

This is to say that managerial actors have usually been conceptualised as if they, by being involved in major technological change projects, were only trying to promote their own managerial objectives. This is true although it can be expected that the action they support must also meet the demands of their personal interests and aspirations. For example, from a perspective emphasising the emotional aspects of managing change, it is evident that in addition to improving skills and advancing careers, the successful overcoming of a technological discontinuity may also foster the permanence of managerial communities and the self-esteem of individual managers. Thus technological discontinuities which necessarily also include problems may be welcomed from various managerial standpoints.

The second section of this chapter therefore examines issues which have been relatively underdeveloped in this field of research and proposes some fruitful avenues forward.

Paths forward in the conceptualisation of management mobilisation

Management mobilisation has been defined, previously, as a temporary social coalition in which conflicting managerial actors temporarily co-operate to bring about discontinuous change. In this section we develop previous conceptualisations of this mobilisation, largely by means of social movement theory. This is because social movement theory has a long tradition in conceptualising the mobilisation phenomenon in its various forms and in various settings. However, although several theorists have welcomed such work (e.g. Pettigrew 1985; Meyer and Zucker 1989; Kanter *et al.* 1992) there have so far been only few systematic applications of social movement theory to management settings (e.g. Zald and Berger 1978; Soeters 1986; Davis and Thompson 1994).

The central assumption of this study is that the knowledge of social movement mobilisation is highly relevant in understanding management. This is mainly because management as a social phenomenon resembles social movements. Accordingly, we believe that analogical reasoning and comparison between social movements and management mobilisation may yield significant new insights to the conceptualisation of the latter (cf. Tsoukas 1991; 1993: 343). The more concrete justifications for such an approach include the fact that management like social movements, creates its power through mobilisation of the resources possessed by separate actors such as personnel or external supporters (McCarthy and Zald 1977). Like social movements, management also represents collective and sustained change action for or against some other social entity or state (Smelser 1963). Even the most outstanding managerial powers may be considered limited in relation to the challenge of initiating and implementing discontinuous technological change. Moreover, like social movements, the forms of management also are dynamic and disposed to change. The management of a corporation has its upswings and downswings during the different phases of corporate development.

However, the most important reason for applying social movement theory to management settings is that social movements are social forms in which conflicting actors are capable of co-operation for a shorter or longer time. It has been argued that this kind of an intra-organisational change force is needed especially in situations of discontinuous change (Zald and Berger 1978). This is because social movements also imply abrupt rejection of previously functional, but recently outdated modes of action and thinking and adoption of new ones. The initiation and implementation of a technological discontinuity clearly fills the criteria for such activity. Managers proposing an ambitious technological change project have to win over several other actors originally sceptical of the appropriateness of such an endeavour. A group of mobilised managerial actors therefore resembles social movements as a proactive and dynamic change force in its specific historical and structural context (cf. e.g. Touraine 1981). Additionally, the social identity of the actors involved gradually changes as a result of the experiences gained (Reicher 1984; Pizzorno 1985) in both management and social movements. For example, the self-confidence of the champions initiating major technological change projects can be expected to reflect their previous successes in similar kinds of endeavours.

It therefore seems clear that previous approaches to management mobilisation will benefit from the systematic use of developments in some related fields of study. In the remaining part of this section we examine three avenues for the further development of the conceptualisation of management mobilisation: managerial actors and their coalitions, the role of resources and the sources of mobilisation decline.

Managerial actors and their coalitions

We argue that the previous literature on management mobilisation has inadequately explored the relationship between individual managers, managerial actors and actors' coalitions. A systematic distinction between these categories means that individual managers may join to form actors and these actors may join to form actors' coalitions. This distinction has two important consequences. First, because coalition-forming in management includes these two separate levels we can distinguish between mobilisation of managers and managerial actors. Second, and more importantly, because both individual managers and managerial actors can make coalitions the number of possible coalitions is multiplied. Consequently, several coalitions may be involved for example in the management of discontinuous technological change. In this book we will show how different coalitions of actors were dominant at the initiation and implementation stages of a technological discontinuity.

Previous writings on management mobilisation thus primarily assume that there is only one change effort at a time going on in management and the firm. Instead, we believe that it is fruitful to assume that instead of one there are many kinds of change programmes going on simultaneously in management. Management should therefore be conceptualised as a constellation of conflicting projects which have distinctive goals and which are promoted by different coalitions of managerial actors. To give an example from paper industry management of the benefits of such a perspective, it can be said that most firms have coalitions of actors which are trying to improve production efficiency through various kinds of processual innovations. Other actors may be simultaneously mobilised to bring about changes in the existing product portfolio. Inevitably, these coalitions of actors end in conflict because of the inherent contradictions between their objectives.

To further illustrate the distinction between coalition forming at the levels of individual managers, managerial actors and actors' coalitions, we have collected some examples from previous studies in Table 3.1. First, a classic example of conceptualisation of coalitions of managers are the managerial cliques (Dalton 1959). They are coalitions which combine the personal interests of different managers (without systematic reference to commonalities in professional background or organisational position) and they often produce negative consequences for the operational efficiency of the managerial hierarchy as a whole. Second, it can also be expected that an individual manager may be a member of several managerial coalitions at the same time (cf. Grieco and Lilja 1996). An example of the conceptualisation of coalitions between managerial actors is the change coalition in the internal corporate venturing model (Burgelman and Sayles 1986). In this conceptualisation actors consisting of individuals with similar kinds of backgrounds (e.g. professional background, organisational position) form coalitions in order to generate a new business activity. Third, an example of the conceptualisation of coalitions between coalitions is the study by Davis and Thompson (1994). It suggests that various coalitions of actors like associations of corporate shareholders on the one hand and corporate management on the other contain internal factions which form new coalitions in the process of competition over the power to control corporate activities. Most importantly, this perspective emphasises that because of the various combinations of different actors within a corporation the outcomes of these separate intentional change efforts always deviate from what was originally expected by their initiators.

Resources as a basis of mobilisation

We argue that previous conceptualisations of management mobilisation have placed too little emphasis on the role of various organisational resources in this process. There are two ways in which organisational resources are important. First, organisational resources are important as concrete facilitators of the change activities. For example, major technological change projects are expensive and often require much planning and other preparation even before the

Table 3.1 Forms of social mobilisation in management

Unit of mobilisation	Type of coalition	Example
Individual managers	Coalitions of managers	Managerial cliques (Dalton 1959)
Managerial actors	Coalitions of actors	Change coalition/ ICV (Burgelman and Sayles 1986)
Actors and coalitions of actors	Coalitions of actors and actors' coalitions	Associations of corporate shareholders (Davis and Thompson 1994)

decision to start the project has been made. Second, organisational resources are important as contextual signs which signal to the actors involved that the technological discontinuity at hand is feasible. This means that organisational resources also influence the ease with which managerial actors are mobilised.

Resources as concrete facilitators of change

Organisational resources form an elementary concept in the contemporary social movement theory. The importance of organisational resources for social movement mobilisation was noted in the 1970s (McCarthy and Zald 1977). Previously, the most common reason for mobilisation to collective action was assumed to be structural strain and shared deprivation (e.g. Smelser 1963, Turner and Killian 1972). In contrast, the current view in social movement theory is that movement activity does not correlate well with grievances and discontent. Incentives are therefore neither necessary nor sufficient for collective action (Davis and Thompson 1994: 152). In other words, collective action provides opportunities for diverse individual and collective benefits and thus it does not require structural strain and deprivation. Most importantly, this notion led social movement researchers to the finding that mobilisation rarely requires conversion of originally disinclined people and might instead involve the transfer of a cohesive group of activists from one location to another (Snow *et al.* 1980).

The advances in social movement theory can therefore be related to a field of research that has recently emerged within management studies and which we have already tentatively discussed above. The new field of research was originally based on the finding that discontinuous organisational change tends to be connected with high managerial turnover (Tushman and Romanelli 1985). Such turnover may be caused by the fact that new actors and their skills and capabilities are needed when new technologies are acquired. The newcomers may be especially effective in this effort if they, for example, can reproduce an already familiar change process. This is not to say that newcomers who lack experience in applying new technology in one setting might not have severe difficulties in applying their knowledge in another (Pennings 1988).

However, in addition to the concrete skills and capabilities, there is also another reason for managerial turnover. The composition of managerial actors may change at the time of discontinuous change because of the commitment of former management to the *status quo*. Instead of forcing and converting the existing managerial actors to act and think in a new way, social mobilisation within management may thus take place through transfers of existing coalitions or networks of managers from outside. This is to say that the process of social conversion may be speeded up through managerial turnover. This is in keeping with the findings that new ideas often enter the management structure through executive succession thus reconstructing managerial beliefs (Child and Smith 1987). In other words, the mobility of groups of managers from one corporation to another may help to break off commitment to earlier traditions and strategies (Grieco and Whipp 1991).

The contextual effects of resources

We accept the general argument that understanding the management of technological discontinuities requires incorporating the context in which management acts into the analysis (Child 1987; Langley and Truax 1994; Wolfe 1994). This is also because internal and external circumstances facilitate and restrict management mobilisation, although they do not determine it (cf. Turner 1981). Resources are one essential element of the context of management mobilisation especially because the amount and quality of resources influences whether managerial actors interpret a technological change project as an opportunity or a threat (Jackson and Dutton 1988). The material and human resources assigned to a major change affect the eventual composition of the project team. A challenging project with limited resources is less credible and attracts fewer new actors than a project with more extensive resources. Moreover, the quality of the resources also makes a difference. A major change project involving corporate partners with good references and individuals with good reputations has better chances to attract more actors to participate.

Temporary changes in material resources (Fennell 1984; Cyert and March 1963: 278–279; Beyer and Trice 1978: 179) and competitive pressures may thus significantly alter the odds for mobilising actors to technological change and innovation. This is not to say that the context of management mobilisation would always be so dynamic. In contrast, some of the contextual characteristics that also can be considered influential resources for mobilisation are relatively static. For example, the geographical location of corporate activities may determine what kinds of issues may act as triggers for mobilisation. It has therefore been argued that the mobilisation of various actors for technological change projects within the Finnish paper industry is facilitated by the well-established positive reputation of these projects (Tainio *et al.* 1989). Such ventures are not only considered promotion of corporate interests but also as collective nationwide efforts to secure utilisation of Finnish wood resources and to maintain the industrial infrastructure.

Decline of management mobilisation

Previous literature has paid limited attention to the reverse side of management mobilisation: mobilisation decline. Whereas much has been written on how people are mobilised, there has been little reflection on why mobilisation fails. This is understandable, considering the traditionally instrumental and practical orientation of management studies. Nevertheless, there is a need to include this aspect in the framework developed here.

It seems that management mobilisation may fail in two senses. This is because both the conversion of the acting management and the transfer of new managerial resources may turn into problems. Failure to convert management has most often been attributed to the inability of the corporate top management to meet the expectations of their subordinates and therefore commit them. As a consequence, creativity and enthusiasm are lost or simply fail to emerge (Walton 1980; Kanter 1983; Burgelman and Sayles 1986). Some other studies on managerial actors have demonstrated that external interference such as the appointment of new heads of operation (Gouldner 1954) or job rotation (Dalton 1959) may jeopardise the well-being of previously cohesive groups of managers.

Extending this existing body of management literature to developments within the social movement theory reveals several similarities. For example, the early theorists in the latter field (e.g. Smelser 1963; Toch 1966; Turner and Killian 1972) noted that social mobilisation is an intensive phase of social action whose life-cycle, however, may be very short. This suggests that the mobilisation of managerial actors is also inherently susceptible to rapid decline and decay. Because the actors' expectations of the consequences of collective action are unrealistic, even the most successful mobilisation will fall short of its objectives. That is to say the objectives of mobilised collectivities are seldom reachable and therefore mobilisation inevitably fails. On the other hand, if the objectives were in fact realistic, their fulfilment would also cause the disintegration of the collectivity. Later this finding was also replicated within management literature; mobilisation decline is caused by escalation of the involved actors' expectations of the joint action (Soeters 1986). In other words, intensive co-operation may easily turn into diverse conflicts within management in situations in which collective action no longer provides benefits for its adherents and supporters (Zald and Berger 1978). Consequently, the enthusiasm involved in even the most successful technological change project will eventually degenerate into routine.

The conversion of managerial actors may also fail because of the contradictions between the objectives of different actors. Because social conversion is often created through alignment of managerial actors' frames of reference (Snow *et al.* 1986) the actual contradictions between the actors involved are blurred. This is the case especially when the announced objectives of the collective action are very abstract and consequently gather together actors from highly heterogeneous backgrounds. Moreover, because mobilisation can also be achieved simply by seducing the actors to take action they do not necessarily support (Ferree and Miller 1985: 52–53) and because such energising is promoted in situations where the individual expectations of the other actors' behaviour change (Klandermans 1984), mobilisation is highly likely to fail. For example, taking part in a major technological change project is motivated by the belief that other actors are also taking part in it, thus making its success more probable. Mobilisation can thus be considered a process in which co-operation creates temporary feelings of consensus and mutual commitment. The basis of this consensus, however, is weak and it most often collapses in the course of action. For example, this study shows how the original enthusiasm gradually weakened during the implementation of a technological discontinuity. Additionally, during its life-cycle the content of action may change and thus produce disappointments and new subdivisions among the actors. The case studied in depth in the following chapters shows how the alterations in the original technology concept disturbed the original initiators of the project.

As far as mobilisation through transfers of managerial actors is concerned it can be argued that replacements of managers at different levels of the corporate hierarchy is problematic, not the least because the entire management structure cannot be replaced. In cases where turnover increases heterogeneity within management, the probability of conflict also increases. For example, state-of-the-art ventures tend to bring in managers with a cosmopolitan identity and the proportion of locals is diminished (Gouldner 1957). The 'successors' may therefore have problems with the 'old guard' (Gouldner 1954: 70–85). In a sense the mobilisation of a new power coalition in management also creates conditions for the mobilisation of its counteraction (cf. Zald and Useem 1987). Thus it is important not to consider the context of management mobilisation as an undifferentiated environment in which the actors left out were unable to fight back. In particular, group level political processes connected to social mobilisation (Hirsch 1990) may lead to polarisation and escalation of conflicts within management.[2] This means that in some cases the opposing groups may avoid open confrontation and try to use their power to adjust the initiated technological discontinuity or delay its implementation.[3]

Conclusions

In summary, in this chapter we have examined the previous approaches to management mobilisation and suggested some promising avenues forward. Both the factors which produce social mobilisation and the factors which impede it have been examined. Some of the most essential contributions on this field are presented in Table 3.2.

On the basis of the material reviewed above, it seems worthwhile to think that factors which mobilise individual managers into alliances and coalitions sharing specific interests and ways of thinking and acting at the same time, also produce new subdivisions within management. In other words, factors which unite managerial actors to form 'us' less evidently also disconnect them from 'the others' as well. Consequently, management can be considered a constellation of actors connected and divided by several factors acting in concert.

The chapter has also proposed that the mobilisation of managerial actors can only be understood in its structural and historical context. Changing situations create recurrent opportunities for the different actors and their coalitions to promote various managerial objectives and interests. As a consequence, we can expect management to be mobilised in situations which allow several mobilising factors to operate simultaneously. Arguing that the mobilisation of managerial actors is situation-dependent does not, however, mean that the number of factors which may lead to this mobilisation is infinite. In contrast, it is assumed that the same factors may produce mobilisation in different situations. For example, we believe that discontinuous change like major technological change projects tend and need to be coupled with social mobilisation in

Table 3.2 Comparison of approaches to social mobilisation in management

| Approach | Features of mobilisation | | | | |
	Period	Author	Content	Source of emergence	Source of disintegration
Indulgence pattern	1954	Gouldner	Social harmony	Flexibility and lenience	External interference
Managerial cliques	1959	Dalton	Promotion of group interests	Emotional bonds	Job rotation
Collective behaviour	1963	Smelser	Collective action to remove an external threat	Structural strain	Tension reduction
Resource mobilisation	1978	Zald and Berger	Internal change action	Individual and collective benefits	End of benefits
High commitment	1980	Walton	Collective action towards ambitious goals	Organic job design	Mechanisation of management
Energising	1983	Kanter	Collective action towards ambitious goals	Empowerment and job rotation	Segmentalisation of management
Collective movements	1986	Soeters	Collective action towards ambitious goals	Socialisation of collectivist values	Failure to reach ambitious goals
Internal corporate venturing	1986	Burgelman and Sayles	Collective change action	Career advancement and empire building	Bureaucratisation

management. This is to say there is, at least, a temporary increase in the intensity of co-operation between different managerial actors representing different managerial levels, functions, professions or cohorts at times of technological discontinuities.

An increase in co-operation in times of technological discontinuities seems to be related to changes in the level of managerial ambition. In addition, there are changes in the composition of management which means that there are new actors bringing in new capabilities and ways of thinking, and thus strengthening the existing management, but also producing changes in the formal managerial structure and the existing alliances and coalitions between managerial actors. From this point of view, management mobilisation implies that managerial actors with different capabilities come into co-operation to produce discontinuous change.

One of the contributions of this volume to the previous management literature is that it further problematises the sources of emergence and disintegration of management mobilisation. It conceptualises managerial actors in an ongoing setting in which their relations to the specific technological discontinuity at hand and the other actors are made explicit. The conceptualisation identifies both the actors' managerial objectives and their personal goals as factors involved in mobilisation. The previous conceptualisations of management mobilisation (e.g. Snow *et al.* 1980) have revealed how networks of previously mobilised actors are transferred from one organisation to another. However, these kinds of formalistic conceptualisations pay no attention to the substantial objectives and goals on which the mobilisation is based. In contrast, this study aims to understand the substantial triggers of mobilisation and the ways these triggers are entwined with each other in a historically specified situation.

In order to justify the arguments presented, the study examines management of technological discontinuities in the Finnish paper industry context. After presenting an overview of these issues at the industry level the study then moves on to examine the mobilisation of managerial actors in the technological discontinuity of Tampella, a Finnish paper industry corporation in the late 1980s.

4 The research process

This chapter describes the research design and data on which the present study is based. The chapter starts by explaining why the Finnish paper industry is relevant as the research object of this study. The chapter goes on to explain why a single case study design was adopted to examine these issues. In particular, it describes why the case selected – the major Finnish corporation Tampella – deserves attention, how the researcher gained access to study it and what kind of data was obtained both on Tampella and on the Finnish paper industry as a whole.

The relevance of the Finnish paper industry for the study of managing technological discontinuities

Although Finland is a small nation, it holds an internationally visible position in the paper industry. For example, in 1995 Finnish paper industry corporations supplied 15 per cent of all the paper products imports in the world and more than 22 per cent of the total imports in Europe. In 1997, the Finnish paper industry companies accounted for 22 per cent of the paper and board production capacity in Western Europe. Moreover, in 1996 the sales of the three largest Finnish corporations made them the fifth, the tenth and the fourteenth largest forest industry companies in the world and the first, the third and the seventh largest in Europe (data derived from the Finnish Forest Industries' Federation and from *Pulp and Paper International* journal). The Finnish paper industry thus plays a critical role in its field throughout the world. The Finnish firms also differ from their competitors in the sense that they refine their output into relatively high value-added products and do not sell large amounts of pulp. Consequently, the Finnish paper industry corporations are not equally strong on the different paper product markets. Instead, their market share is high especially in writing and printing papers such as fine papers and coated printing papers. The market shares of Finnish paper industry corporations in the wood-containing printing paper grades shown in Figure 4.1 clearly illustrate the Finnish dominance in these areas. Moreover, during recent decades the Finnish paper industry corporations have also expanded internationally. Currently, they have production units in various locations in Central and Western Europe (e.g. Germany, France, Britain and Spain), in North America and most recently, in East Asia.

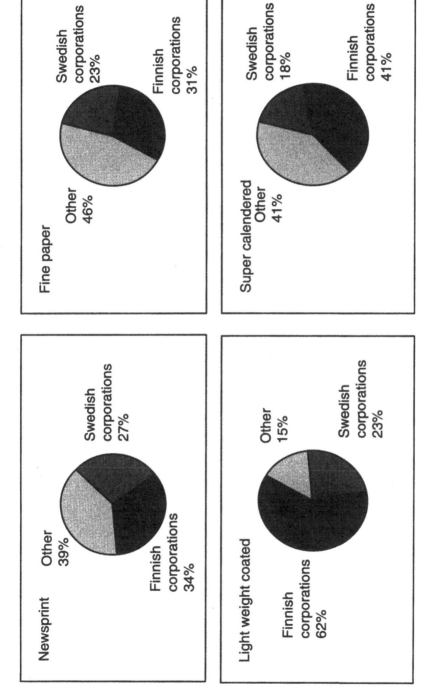

Figure 4.1 European paper production capacity in printing and writing paper grades in 1997
Source: Finnish Forest Industries' Federation.

Interestingly in relation to the theme of this book, the Finnish paper industry has improved its position especially through the systematic modernisation of its production technologies. More concretely, the transformation of the Finnish paper industry is based on a long-term process in which overall technological development has been converted to numerous major technological change projects in different corporations and different mills (Rohweder 1993). This means that the Finnish paper industry corporations have become capable of managing greater and greater production capacities in the new facilities. For example, since the 1980s the Finnish paper industry corporations have had the largest average production facilities (i.e. paper machines) in the world. They also have increased production efficiency by improving technological integration between different production units in the vertically integrated production chain. Simultaneously, continuing development of existing facilities has permitted gradual increases in their production output.

The expansion of the Finnish paper industry corporations has thus been built largely on the successful management of technological discontinuities. A large number of change projects aimed either at improving efficiency in producing a specific paper grade or at launching a totally new paper product (either for the mill, company or the industry as a whole) have been initiated and implemented. This has been the case although the scale of the new technologies (e.g. paper machines) has increased steadily. The challenging nature of the transformation of the Finnish paper industry is partly related to the fact that growth and an important market position has been achieved mainly through building new production facilities and renovating the old ones. In contrast, increasing market share through acquisitions of other paper producing companies from abroad have been rare.

One part of managing the technological discontinuities has been the successful cloning of change projects both within and outside Finland.[1] The typical situation in the latter case has been that the Finns start to produce relatively unadvanced paper products (e.g. newsprint) in a foreign country with technology that is (at least in principle) similar to the technology already in use in the domestic facilities (see e.g. Laurila and Gyursanszky 1998). An important prerequisite for successful transformation has been the existence of competent domestic suppliers of new technology and technical consultants. As a result, these firms, which operate in the field of technology design, construction and project management, have become leading figures in their field of expertise. Finland therefore has Jaakko Pöyry, the leading forest industry consultancy firm, and Valmet, one of the three major international paper machine manufacturing corporations which has built its success largely on intensive co-operation with the Finnish paper producing companies (see Alajoutsijärvi 1996).

The aim is not to contend that the Finnish paper industry corporations do not have any problems. First, the dominance of the forest industry products (which is illustrated in Figure 4.2) could be considered negative for the Finnish economy as a whole. A more concrete problem and probably one of the negative side effects of the numerous technological change projects, is that the profitability of the Finnish firms has not increased in pace with production efficiency (see e.g. Pohjola 1996). It should be acknowledged that this is partly due to heavy investments in new technology. As a result of these investments, the Finnish paper industry corporations have borrowed heavily and become highly dependent on their sources of financing. The long-term capital structures of the paper industries in Finland, Sweden, Canada and the United States are compared in Table 4.1.[2] Thus technological discontinuities have necessitated external resources which have momentarily interrupted cash flow. This is both because new production technologies have often been adopted before they were systematically tested elsewhere and because some of the old technology that has been dismantled might still have been used for some time.

Another problem is that the Finnish paper industry companies have not been especially innovative in developing new products. Instead, in most cases, they have developed new production methods for products that others have invented. Although most of these products represent relatively high added value, they often turn into bulk products as the size of the market grows. The relatively low rate of research and development expenditure and the highly concentrated product portfolios are a result of the fact that most of the technological change projects have expanded the production of paper grades that the firms were already manufacturing.[3] As a result of operating mainly in the fields of standardised products, the profitability of the paper industry corporations has remained vulnerable to fluctuation in the prices of raw materials (e.g. wood, energy). Moreover, in times of recession, imports from the underdeveloped paper producing countries pose an additional threat.

The most recent developments in the Finnish paper industry corporations include modifications in corporate policies aimed especially at improving the image of the industry among international institutional investors and environmental pressure groups (see Halme 1997). The Finnish companies are therefore

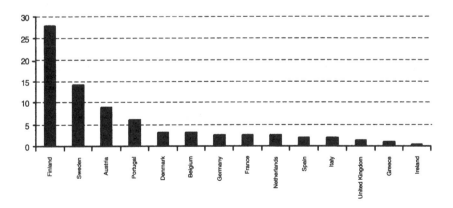

Figure 4.2 Forest industry exports as a percentage of total exports in different EU countries in 1996

Source: Finnish Forest Industries' Federation.

Table 4.1 The long-term capital structures based on the market values and cash flows, 1982–1993, in the paper industry

	Finland	Sweden	Canada	USA
Equity (%)	22	37	40	63
Debt (%)	78	63	60	37
Total capital (%)	100	100	100	100

Source: Adopted from Artto (1996: 79), reprinted by permission of the *Finnish Journal of Business Economics*.

expanding their operations not only by building new mills but also by acquiring existing mills both in Europe and the United States. The companies have also been active in creating a more systematic brand strategy for their products and in being innovative in various kinds of product architectural innovations (Henderson and Clark 1990). This means combining the already existing paper products in a new way to create new products. As far as the general management of the corporations is concerned, it seems that the new tendencies also imply a shift from the previous emphasis on the investment decisions and a financial control type of management style (Goold and Campbell 1987) of the corporate activities to a somewhat more flexible and creative approach. This implies that the customer's perspective becomes more important and also that non-line management experience becomes valued. Consequently, experts in staff positions obtain more power.

However, there are concrete obstacles to radical change in the management of the Finnish paper industry corporations. This is because creating corporate competitiveness through technological advancement will also remain a key issue for the paper industry in the future. Moreover, the paper industry (at least in Finland) is dominated by engineers and the self-confidence of both the management and personnel is largely based on the modernity of the production facilities. Moreover, managerial competence has customarily been tested in major technological change projects. The fact that it is highly difficult to promote technological modernity and a cash-flow based management orientation within the same company also inhibits change. For example, smooth co-operation between personnel producing high quality and bulk products within the same mill cannot be expected. As a consequence, the Finnish paper industry is and probably will remain an interesting context for studying the management of technological discontinuities. The next section explains why these issues are studied in a longitudinal case analysis within the Finnish paper industry.

The argument for a longitudinal single case study

This book includes a longitudinal case analysis of the mobilisation of managerial actors to manage technological discontinuity. This analysis is based on case study evidence from the Finnish paper industry firm Tampella. It focuses especially on a single major technological change project at Tampella's paper mill in the late 1980s. It is acknowledged that a study concentrating on one company in a specific business context is always limited with respect to statistical generalisability. Nevertheless, the case study approach is totally adequate here because of the opportunities it creates for observing and describing a complicated research phenomenon in a way that allows analytical generalisations (Dyer and Wilkins 1991; Eisenhardt 1989; Tsoukas 1989). Several classical studies in management research have demonstrated the usefulness of focused case studies in the examination of longitudinal change processes (e.g. Pettigrew 1985). This is partly because such a research design permits consideration of the specificities of the business context (Child 1988). Consequently, the most important factors justifying a one-case case study are the selection of a representative or revelatory case (Yin 1989) and the use of appropriate research methodology. Although these issues were already taken up in the Introduction, they will now be discussed at greater length in this and the following section.

Tampella was selected for this study partly because it had managed technological discontinuities in its recent past. More concretely, it had carried out major technological change projects at its paper mill just before the field work for this study began. The technological discontinuities at Tampella entailed introduction of new technology and dismantling of old technology which was the requirement for the replacement of some of the previous paper products with new ones. Thus the Tampella case is specifically about managing technological discontinuities. This is not, however, anything exceptional. The other Finnish paper industry companies also carried out such projects and many of them took place before the project in focus here. The revelatory nature of the Tampella case is therefore not related to what actually was done but instead to the relationship between what was done and what had been done before. In other words, what is interesting here is that suddenly Tampella, which had not been well-known as a representative of the most advanced paper products and production technologies, initiates and implements a major technological change project aiming at the state of the art in paper making. We believe that examination of the process whereby a previous technological laggard tries to catch up with the technological leading edge offers considerable potential for conceptualising how technological discontinuities are managed.

We would therefore argue that the case of Tampella is interesting especially because the company completed an ambitious change project with relatively few resources. Most essentially, the project in question here implied that Tampella, which had previously focused on the production of technologically non-demanding paper grades, tried to become one of the leading printing paper manufacturers by adopting new on-line coating technology which had only recently been developed. One of Tampella's problems concerning paper industry development was that its competences were in a relatively large area within the forest industry sector. The company produced paper and board and also the machinery needed for their manufacture. It had only one mill producing printing papers whereas its other facilities produced quite different types of paper products or paper machinery. This means that Tampella initiated and implemented a major technological change project at its paper mill although it had – for a Finnish paper industry corporation – exceptionally little recent experience of such projects. In fact, Tampella carried out two such projects during the 1980s, although here we concentrate on the latter one because although the first project significantly increased Tampella's paper production capacity, it did not incorporate a major discontinuity in the line of production (see Laurila 1997a).

In sum, the Tampella case offers an opportunity to examine general problems in the management of technological discontinuities in an exceptional setting. The investments in new technology at its paper mill accounted for 20 per cent of Tampella's total capital outlays over a ten-year period. These and some other details of Tampella's performance are given in Table 4.2. These investments were therefore a significant issue on the different management levels of the firm and highly visible to all managerial actors. As a matter of fact, in the 1980s the paper mill was Tampella's largest business unit with a yearly turnover of between US$200 and 300 million and a 15 per cent share of Tampella's operations. As a consequence, the timing and content of the change activities could easily be identified and managers were aware of the background of each project. But the mere identification of an interesting case does not guarantee adequate research data. Therefore, the field-study methods and evidence obtained in the study are now described.

Research access and evidence obtained

This book is based on archival and oral history evidence despite the difficulties involved in this kind of retrospective interpretation (cf. e.g. Barley 1990b: 228; Schwenk 1985; Golden, B.R. 1992).[4] The longitudinal study of Tampella consists of a retrospective analysis of the development of a paper-industry firm and its paper mill through numerous written documents and in-depth interviews of the key actors. The access to this material was obtained by approaching Tampella through multiple ports-of-entry. First, before starting this research I had already completed one study (Laurila 1989) in which Tampella was used as an empirical example. Consequently, I was rather well informed of what was happening at Tampella and also had contacts with Tampella managers. Second, after finishing the research proposal for this study in the spring of 1990 I contacted managers both at corporate level (a member of the Tampella board) and at division (the head of Tampella forest industry division) and business unit (the head of the paper mill and a production manager) levels. Soon thereafter the research proposal was accepted in principle and no attempt was made to direct or restrict the field work. In other words, while the field work received official approval, no goals or direction were imposed upon the researcher's task.

Third, in addition to being an individual researcher, I was at the same time a member of a research group making a more general survey on production technology at six paper mills in five Finnish paper industry firms, of which the Tampella mill was only one. Membership in this group enhanced access, which

Table 4.2 Tampella, 1980–1990

	1980	1981	1982	1983	1984	1985	1986	1987	1988	1989	1990
Turnover (US$, million)	428	574	555	548	830	898	863	944	1247	1520	1710
Total investment	49	73	213	152	88	75	66	105	311	561	368
Operating margin (%)	14	18	10	7	15	16	12	13	14	9	6
ROI (%)	8	15	4	1	9	11	7	9	12	8	2
Debt per turnover (%)	93	81	121	147	125	116	117	120	105	134	127
Share capital	15	23	23	23	23	23	34	43	65	73	75
The paper mill's share of turnover (%)	14	12	11	14	17	20	21	22	18	13	12
The paper mill's share of investment (%)	15	54	61	63	11	20	16	11	14	19	3

Adopted from Laurila (1997a: 223), reprinted by permission of Blackwell Publishers Ltd.

is always critical in research based mainly on one case (Leonard-Barton 1990: 263) by providing me with evidence I could not have obtained from the Tampella case alone. Instead of having information on only one facility, I had access to material concerning five other units in four separate firms. Among other things, the evidence from the five other mills facilitated comparison of the timing and content of accomplished technological change projects in these companies. This was essential for the analysis presented here because only through comparison with other mills within the same product area is it possible to assess the technological ambitiousness of a specific project. Finally, in addition to being a member of an established research project, I also had contacts with several other management scholars who did research at Tampella. These contacts (e.g. Ahlfors 1993; Vaara 1993; Saarinen 1994) permitted comparison of results and experiences concerning the various features of Tampella.

The multiple contacts described above provided me with adequate research access to Tampella. The fact that there was no systematic control by the company managers on my work helped me to avoid the potential disadvantage of being led to study specific issues at the cost of neglecting other more relevant ones (Dalton 1959: 275). This is not to say that there were no expectations from my work on the part of Tampella managers. At every stage I made it clear that my study was focused on the relationship between management and technological change. My belief was then (and still is) that management will not allow access at all if they do not understand on their own terms what is being studied (see Jackall 1988: 13–14). Managers may deny access if they are unaware of the kind of data the researchers are looking for and the purposes for which the data will be used (cf. Coleman 1996). Thus unrestricted access is seldom granted, and researchers instead need to utilise empirically defined research topics through their own conceptual apparatuses. This was possible here because neither the methodologies nor conceptual frameworks used in the study were restricted.

After describing the quality of the access, we should then consider what kind of data was obtained with such research access. Roughly speaking, the data obtained can be divided into two categories: written documents and oral history. The data on the Finnish paper industry include information collected from public data banks (for example the data bank of the Finnish Forest Industries Association), professional paper industry journals and a variety of both professional and academic research reports. The documents concerning Tampella include internal memos and newsletters, production statistics, annual reports, business histories, published biographies, organisational charts and published articles. More detailed information on the contents of the internal memos has been provided in Laurila (1997a: 224). The comprehensive list of collected articles can be found in Laurila (1995: 149–156).

Production statistics include Tampella's production volumes of the different paper grades and outlays on paper production technology between 1980 and 1992. Some data were also obtained concerning changes in the level of production efficiency during the same period. Tampella annual reports provided details on the profitability and content of the corporate activities. These documents were available from the year 1960 onwards. The published business histories (e.g. Lodenius 1908; von Bonsdorff 1956; Teerisuo 1972; Seppälä 1981) and personal biographies of Tampella managers or board members (Alho 1961; Björklund 1982; Ehrnrooth 1991; Saari 1992) created a basis for describing the long-term development of Tampella both as a technical and an entrepreneurial activity. Organisational charts were used to identify the individuals involved in the different phases of corporate development.

In the case of Tampella the number of published articles both before and after the turn of the 1990s is vast. The list of articles collected for this study was mentioned above. The number of articles is high as a result of the ambitious changes at Tampella at that time. The changes made the news especially because they so clearly contradicted the previous reputation of the company and because Tampella corporate management intentionally used publicity as a tool for promoting the changes it had initiated (see Laurila 1997c). This means that in a sense discontinuous change or even corporate crises (Useem 1995: 30) may facilitate research access through making researchers valuable to managers. However, the fact that there were lots of published writings on Tampella does not mean that such material should not be treated cautiously as research evidence. Nevertheless, it seems justified to claim that in this case they provided the researcher with at least a starting point in conceptualising the mobilisation of managerial actors in the situation of discontinuous change. This is because the published material included numerous details on the acts and views of individual managers at the different management levels in the different phases of the change process.

Technical details on the technological discontinuities at Tampella were obtained mainly from archival data. However, to provide a managerial perspective on these technological discontinuities, the author formally interviewed fifteen managers in key positions at Tampella at the time of the change projects. Details of the interviews are given in Table 4.3. The material obtained in the interviews was used especially to incorporate the perspectives of the various actors involved. It is important to note that although the number of interviews is not high, the range of interviewees adequately covers the different management levels involved in the process. Most of the interviewees were managers who had closely observed the change projects at the Tampella paper mill during the 1980s. These individuals were interviewed from one to three times in 1990–1994. The interviews were open-ended and semi-structured and they lasted from one to three hours. These discussions concentrated on the details of the change processes and each actor's actions and aspirations and the perceptions of others in the same process. To encourage confidentiality and the production of information the interviews were not tape-recorded. All the handwritten memos of the interviews were typed immediately afterwards and their total length amounted to 175 double-spaced pages.

The essential role of interviews in the reconstruction of the processes involved in the management of the technological discontinuities at Tampella may jeopardise the validity of the research findings. This is simply because the

Table 4.3 The interviewees and their management position

Job in the early 1980s	Job in the late 1980s	Number of interviews
1 Chief Financial Officer	CEO (at the beginning)	2
2 –	Chairman	1
3 –	CEO (at the end)	2
4 Head of Forest Industry	Starting to head Business Development	1
5 –	The new head of Forest Industry	1 (also discussions by phone)
6 Head of Machinery Division	Head of Machinery Division	1 (by letter)
7 –	Head of the paper mill (at the end)	1 (also discussions by phone)
8 –	Head of the paper mill (at the beginning)	1 (also informal discussions)
9 Technical manager	Production Director	3 (also discussions by phone)
10 –	Marketing Director	2
11 –	Sales Manager	1
12 –	Head of Production Development	1
13 Business manager	Head of Management Development	2
14 Production supervisor	Chief Instructor	1
15 Production manager	Project Manager	1 (only by phone)

Adopted from Laurila (1997a: 225), reprinted by permission of Blackwell Publishers Ltd.

motivation of the informants to co-operate with the researcher is a necessary but not a sufficient condition for obtaining valid information. The interview situation should therefore be considered reciprocal in the sense that managers need to get something in return for their time and energy (Barley 1990b: 238–239). Moreover, in cases where the researcher is clearly not an insider in the case organisation, it is obvious that the informants may – if it serves their interests – give distorted or even false information. More probably, instead of giving adequate answers the individual informants may simply reproduce prepared speeches (Thomas 1993) as a response to the researcher's questions. One part of the problem is that formal research access does not guarantee the rapport of the individual informants. The least that researchers therefore need to do is to translate their research ideas again and again to the individual informants (Hirsch 1995) or in some way 'break the ice' to build useful contacts. They must also ensure the relevance of the material produced.

Although these problems may not have been the most relevant in this study – the researcher was familiar with the Tampella case from the beginning of the research process – an account of the interview process in this study seems pertinent.

Managing the interview process

I believe that in the actual interview situation there is no single way to secure a co-operative relationship between informant and researcher. However, it seems evident that the way the contact is formed affects the quality of the evidence gained from the informants (Buchanan 1993). Thus when working in the field I emphasised my independence of any interest group involved with Tampella and reminded the informants of my aim to seek multiple perspectives to the change process. This way I sought to create the 'social space' necessary for discussion of the events. This was especially important because it can be expected that research where the managers themselves are in the focus of the study is both ambiguous and emotionally charged.

It can be assumed that in general, research is a marginal activity for most managers. This is because managers often consider the concepts and categories of researchers theoretical and not useful for practical purposes (Hirsch 1995: 76–77). Researchers therefore need to do all they can to arouse personal interest in the study before the interview. I personally did this by submitting questions beforehand and formulating the questions in a way that demonstrated their relevance to the managers. To achieve this, I had gathered information on each manager's background and experience at Tampella. In addition, I had written a description of the main features of the Tampella paper mill before the first formal interview for this study in June 1990. A little later this description was extended to include the principal technological and managerial events. Through descriptions and previous reports (Laurila 1989; Laurila 1992) I tried to help the informants to recollect and produce their own interpretations of the events and their own role and that of the other managers in them.

However, it can be argued that no material would be enough to ensure recollection if the events themselves were not meaningful to the informants. It can be argued that the technological discontinuity in focus is one of the most important phases in the working history of the informants. We can therefore expect that they will readily recall even those events that occurred several years earlier (cf. Golden, B.R. 1992:849). As an indication of awakened interest, most of the interviewees accepted the proposal for an interview without hesitation and all accepted it after receiving a more thorough account of the study. Moreover, soon after the meeting, I sent a memo on the interview to many of the informants. In some cases this provided essential new information because the informants wished to sharpen their arguments. It also expressed the emotional tension connected to me and the issues under examination. Furthermore, correspondence concerning the memos of the interview provided me with knowledge of the informant's further willingness to co-operate.

It can thus be argued that the interviews conducted for this study were of high quality. I was well prepared before the interviews, most of the informants were very co-operative and second or even third interviews and written and oral feedback facilitated in-depth examination of the issues in focus. Nevertheless, it must be acknowledged that the collected evidence is not in any sense complete or conclusive. Although the key figures involved in the processes were interviewed once or twice, many more interviews could have been conducted. Moreover, one problem was that not all of the interviews produced much information on processes of mobilisation (see Laurila 1997c). We therefore examine below the ways in which the inevitable deficiencies in the interview evidence in this case could be overcome.

The first way to supplement and validate the interview evidence was by using the documentary evidence to cross-check many of the details presented in the interviews. It was possible to control the reliability of the managers' recollections on technical and other details by comparing them with information in the documents. This means that the number of unintentional errors of recollection was restricted through triangulation (Jick 1979). Second, in addition to interviews there were also informal discussions with some thirty-five managers and consultants who had been employed or otherwise closely involved with Tampella. Five of them were acting managers at the different management levels. In addition to the fifteen interviews already mentioned I formally interviewed the head of a nearby paper mill – one of Tampella's closest competitors – who knew many of Tampella's managers personally. Additionally, I also interviewed a manager who had headed another competing mill and who at the time of the interview had headed Tampella paper mill for six months. Moreover, although the research was not ethnographic in any systematic sense, I also had some opportunities to make informal observations of the managers and other Tampella people in their work settings. For example, I visited the mill shop floor several times, spoke with managers in their offices, over lunch and in a company limousine and spent days perusing files at various Tampella offices.

Moreover, twenty-seven managers employed by Tampella's paper industry competitors commented on my reports on Tampella either orally or in writing. Obtaining feedback was partly facilitated by the fact that during the research process the Tampella case was constantly referred to by its own name which permitted response from a wide public including researchers, consultants and practitioners. The wide variety of feedback to drafts and oral presentations provided significant new information. More specifically, they challenged previous interpretations concerning Tampella and also provided additional information on the case, the individual managers and their actions.

However, we do not argue that this study is unique because of the large amount of evidence gained. The research process of this study is exceptional in the sense that the researcher had the opportunity to collect and assess the evidence and to learn about the microcosm of Tampella and the Finnish paper industry over a five-year period. The different research activities are illustrated in Figure 4.3. Moreover, since the field work on Tampella ended I have continued to do research (e.g. Laurila and Gyursanszky 1998) in paper industry settings and have been in contact with paper industry managers on various occasions. This is important merely because it has permitted the gradual development of various research hypotheses and the adoption of new theoretical perspectives on the issues in focus during the research process. In the work on Tampella it was also possible to compare the interview evidence with evidence published during the research process and to make use of the published evidence in later interviews and other contacts with the informants. Moreover, the long-term research effort permitted use of previous publications to gain more information and to facilitate examination of delicate matters such as the role of individual managers in managing technological discontinuities. For example, it was possible to get into contact with individual informants at a time when they could be expected to be most motivated to give their personal account of the events. Some of the Tampella managers in this study were therefore interviewed after they had already left the company. Although not necessarily always the case, my experience was that this allowed them to evaluate the issues more freely.

Taking a more general perspective, it seems justified to argue that the exceptionality of studies like this is essentially related to the fact that the researcher is able to conceptualise the interaction within management through multiple contacts to the field and at the same time to hold a non-participant position in the social arena of management. In such a case, instead of merely making observations of a phenomenon which exists irrespectively of the researcher's own existence, the researcher operates in its very sphere of influence. Thus the fact that I observed managers and discussed with them in different situations and at different moments in time brought some evidence on the effect of different events into the relationships within management. For example, I interviewed some managers while they were working for Tampella and after they had left the corporation. Thus the critical incidents which both the company and its managers faced during the research process made it possible to conceptualise the setting-specific cues (Barley 1990b) of social mobilisation in the Tampella case.

To conclude, as a result of the above described research process I have

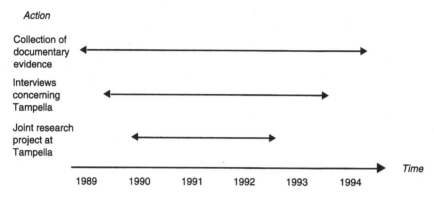

Figure 4.3 Time action plan of the research process

completed a study which continues the previous research traditions on managerial actors but which also makes a specific contribution to that literature. One of the specificities of the study is that it is one of the few analyses (e.g. Burawoy and Hendley 1992) in which the changing objectives and relations of conflicting managerial actors have been identified in an openly examined historical context. Compared with the previous studies exploring managerial action in a specified corporate context (e.g. Pettigrew 1987; Burgelman 1991; Burgelman 1994), this study makes a contribution in the sense that it more explicitly problematises the relations between different managerial actors. This book has still another distinctive feature. It examines managerial action in relation to concretely specified change: technological discontinuities. To achieve this, the acts and features of individual managers have first been examined. However, the acts of individual managers are in this report explored only to the extent necessary to provide the reader with an overview of what actually happened in the management of the technological discontinuities. In fact, the individual managers are not interesting in themselves, but instead as parts of the managerial actors they collectively form. Thus the managerial actors on which this study focuses are aggregations of individual managers and as a consequence, the features of the individual managers fade. Therefore the details of the management of technological discontinuities in Tampella are provided in Chapter 5, and the aggregated and idealised actor construction and analysis in Chapter 6.

After having described the research process the study moves forward to the detailed case description and analysis. It starts with an overview of the technological discontinuities at the Tampella paper mill, which aims to provide an understanding of the conditions and specificities of the discontinuity and the action gap it created in relation to the mill's previous experience.

5 Overview of the technological discontinuity in focus

This chapter provides an overview of the setting in which the management of a specific technological discontinuity is examined. It therefore analyses the features of the products, production technology and management of Tampella paper mill before, during and after the specific technological discontinuity in focus. In the following we will briefly describe the development which led to the relative decline of Tampella's paper industry and the situation of 'do or die' in the late 1980s in which a major technological change project was initiated and implemented. In other words, after the conditions for the discontinuity at Tampella have been described the analysis then moves on to the specificities of the initiation and implementation of the major technological change project. The major reason for describing and analysing these details is to offer the reader a foundation for confronting the more abstract managerial actor perspective applied to the same technological discontinuity in the next chapter.

The Tampella paper mill before the discontinuity

This study on the management of technological discontinuities focuses on technological changes in the late 1980s at the Tampella paper mill, founded in the late 1930s. In this section we analyse Tampella's previous technological standard and managerial tradition in relation to the discontinuity in the late 1980s. In particular, it will be argued that Tampella paper mill was technologically outdated and that it had few managers with 'up-to-date' skills and capabilities in paper industry technology before the discontinuity. This is to say the mill management lacked knowledge of sophisticated paper products and paper production processes and of their management as a business.

Gradually losing contact with state-of-the-art technology

The Tampella paper mill started operations in 1938. At that time the mill – with its up-to-date paper machines and necessary pulping facilities – was a relatively large and technologically advanced production unit. However, production technology at the mill between the late 1930s and the late 1980s had become increasingly obsolete. This is because Tampella gradually lost contact with the leading edge in paper making technology. This process of decline comprises four critical incidents, which will be examined below.

The first critical incident in the process was the fact that the start-up of Tampella paper mill was too late to benefit from the original boom on the international newsprint market. In fact, Tampella started to make plans for paper production in 1935, when the demand for newsprint was rising sharply. The decision to build the mill was made in 1936. However, newsprint prices were already falling in 1937 and it was not until late 1939 that the new mill reached full production. The worst was yet to come, however, because war broke out soon after the mill started up. As a consequence, the mill was able to produce only one-third of its capacity during the years 1939–1944. The post-war shortage of energy also delayed achievement of full production until 1949. Thus the Tampella paper mill lost a great deal of its potential production during the first ten years of its operation because of external circumstances. In a way, Tampella also lost the time needed to accumulate wealth needed for further modernisation. Nevertheless, during the 1950s the mill was still one of the most efficient producers of newsprint on the market because of its large machines, limited number of paper grades and large average order size.

The second critical incident in the development of Tampella paper mill was the partial turn from the production of newsprint to the production of newsprint speciality grades since the beginning of the 1960s. This was because Tampella's paper production technology had gradually fallen behind that of its competitors. At that time the two other newsprint mills located nearby, which had also started their operations in the 1930s, extended their operations by building new paper machines. To compensate for the lack of new machinery, Tampella turned to the production of slightly more refined speciality grades which had been developed a few years before. However, because there were no major investments made or experimental equipment involved in the introduction of the new grades, productivity started to decline. The price of speciality grades was only somewhat higher than that of newsprint grades. This was because the production of speciality grades was not difficult enough to prevent the newsprint standard grade producers from entering the speciality grades market at times when the price difference was adequate.

The third critical incident in this gradual decline took place in the late 1960s. The lack of competitivity of the Tampella paper mill became obvious. In fact, the shift to the newsprint speciality grades had proved to be only a delaying action. The profitability of the mill remained low because of numerous variations in the speciality grades and a low average order size. In addition, the production machinery of the mill was gradually becoming obsolete. In 1969, plans for major investments in production (including a new paper machine and corresponding extensions and modernisations of the pulping facilities) were made but they were frustrated by a lack of funds caused by the company's international acquisitions in North America.

As a consequence, in the 1970s the Tampella paper mill was technologically outdated when compared with its domestic competitors. It may be noted that it was the only Finnish paper mill (out of thirty) which in its core production technology (paper machines and main pulping facilities) relied solely on technology from as far back as the 1930s. In the words of the mill management at that time 'it was completely impossible to make a profit with the existing machinery' (*Insinööriuutiset* 16 October 1985: 8). Tampella thus left its paper mill to survive with its existing technology for another ten years. During that time the mill produced both newsprint and newsprint speciality grades and also book paper grades which the mill had adopted to its product portfolio through minor adjustments in the production process. Thus Tampella had a flexible paper mill; it had small capacity but it produced a large variety of printing papers with low value-added.

The fourth critical incident in the process comprised the major technological change projects performed at the mill in the early 1980s. The project was a revised version of the one rejected some ten years earlier. More concretely, between 1981 and 1983 a new mechanical pulp mill and a new newsprint machine (PM 3) were built alongside the two earlier machines (PM 1 and PM 2). Unfortunately, the changes failed in many ways. The decision to expand newsprint production could be justified by the prospects for newsprint demand which were still good in 1980. However, the fact that the Tampella paper mill used these investments to resume large-scale newsprint production which it had gradually given up during the previous decades aroused suspicion. These suspicions proved to be valid as the markets were saturated before the actual start-up two and a half years later.

It can therefore be argued that building a new paper machine did not stop the long-term relative technological decline of Tampella mill. Partly because of the market saturation, it took four years to reach an efficient production level and to make the mill profitable. Tampella's position was especially weak because the new paper machine happened to be the last newsprint machine out of the three started in Finland in the early 1980s. The production of the new machines (as well as the old ones) at that time was all sold by the joint sales association Finnpap. Because Tampella's new paper machine was now the latecomer among the newsprint producers, the company received many of the technically inconvenient orders which made it even more difficult to increase the efficiency of production.

In summary, the original start-up of the Tampella paper mill was followed by a gradual comparative decline aggravated by shortcomings at a few critical junctures. The main characteristics of these critical incidents are presented in Table 5.1. First, the start of the new paper mill was followed by years of war and gradual loss of competitivity. Second, the shift to newsprint speciality grades partly enhanced this process. Third, the timing of the awaited modernisations in the early 1980s was not successful and in particular, the paper grade chosen for the new paper machine had become too unsophisticated for a mill located far from the main markets. As a result of this process, the Tampella paper mill in 1987 was a technological laggard compared with its domestic competitors before the technological discontinuity in the focus of this study. This is to say

Table 5.1 Critical incidents in the decline process of the Tampella paper mill

Timing	Background	Content	Main problem	Effect
Late 1930s	Increase in newsprint demand	Building of the paper mill	Timing	Missing of the best period
Early 1960s	Investments by competitors	Moving to specialty grades	Increase in the number of grades	Decrease in production efficiency
Late 1960s	Technological obsolescence	Investment proposal	Lack of financial resources	Postponing investment
Early 1980s	Technological obsolescence	Investment project	Outdated product	No change in the competence level

that Tampella in 1987 was still one of the least advanced printing paper producers in Finland.

Tampella's relative technological backwardness is indicated by several features. For example, Tampella's paper machines one (PM 1) and two (PM 2) ran only at 750 metres per minute, whereas the average speed of the Finnish printing paper machines at that time had already exceeded 1,000 metres per minute (Jaakko Pöyry 1987). Apart from the fact that the technological standard of the production equipment was relatively low, the operations were not profitable. Though prices for newsprint had risen they did not permit profitable production. In fact, the price of newsprint in the middle of the 1980s was 20 per cent below the level anticipated at the time of the new paper machine investment decision in 1980. At the same time many domestic competitors of Tampella had given up newsprint production. This tendency is illustrated in Figure 5.1. Tampella competitors had replaced newsprint with more refined paper grades such as SC (super calendered) or LWC (light weight coated). Among other things, such conversions necessitated adding of calendering and coating processes to the technology used in the production of newsprint. This means that Tampella had acquired handicaps by not learning many of the new capabilities common among its domestic competitors.

After showing that Tampella was technologically in a rather weak position for initiating and implementing a major technological discontinuity, it is interesting to examine what can be said about its management in the same field. It is particularly interesting to assess what kind of management tradition had been formed in the long-term decline process and what was the level of managerial competence in the late 1980s. The following subsection therefore analyses the major features of the Tampella's paper industry management tradition before the technological discontinuities in focus.

%

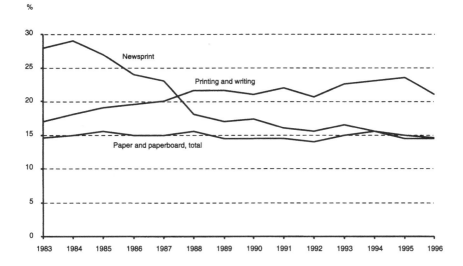

Figure 5.1 Finland's share of paper production in Western Europe
Source: Finnish Forest Industries' Federation.

From patriarchal production management towards business management

Tampella's paper industry management tradition (like that of the other Finnish paper industry companies) can be termed patriarchal production management. Features characteristic of that tradition include an underdeveloped division of labour within management and a concentration of power in a few key managers. The patriarchal tradition implies that the managers are strongly committed to the mill and the company and that recruitment of managers from outside and also the lateral mobility of managers inside the company is rare. Moreover, such management is dominated by emphasis on the smooth flow of day to day production and the tendency to continue in the chosen lines of business.

The key figure in whom the patriarchal tradition was personified at Tampella was Bert Brofelt, the engineer who headed the Tampella paper mill for thirty years from its initial start-up until the end of 1968.[1] Because Brofelt was originally one of the few at Tampella with earlier experience of paper production, his professional authority since the start-up was obvious. During his era Brofelt was the father figure responsible for everything that happened at the mill and also for much that took place in the local community. His management style was personal and emphasised modesty. Successes were not celebrated loudly, and attention was paid to even small things like greeting each other and turning off the lights at the mill. Brofelt insisted that Tampella was in no way weaker than its competitors.

During the 1960s the authority of the patriarchal management faced increasing threats. For example, the personnel at Tampella mill had expected the building of a new paper machine since the beginning of the 1960s. As this did not happen the authority of the mill manager was needed to persuade the other managers and personnel to support introduction of newsprint speciality grades. In a way this was something to do while waiting for the new paper machine. In the late 1960s Brofelt retired and Kaarlo Ahopelto, the former technical manager, was appointed the new mill manager. Before his retirement, however, Brofelt as a member of the board of Tampella, had started planning for the new paper machine. Ahopelto was responsible for accomplishing these plans, which (as mentioned above) were soon cancelled. The cancellation was a great disappointment for the management and personnel at the different levels of the hierarchy. The personnel felt that 'they were not worth the investment' and that 'Tampella would probably sell its paper mill to some other company'.

The fact that no major investments were made at Tampella for almost fifty years ruined the ideology behind the patriarchal management tradition. It was difficult to believe that Tampella was among the top paper producers when even managers like Kurt Dahlvik, the head of Tampella paper mill between 1973 and 1982, later confessed that at that time 'the mill seemed dilapidated' (*Metsäteollisuusuutiset* May 1989: 4). However, both Ahopelto and Dahlvik continued Tampella's previous management tradition in the sense that they had both made a long career in the company and that when working they intervened in virtually everything going on under their command. They were also jointly responsible for management of the technological change projects in the early 1980s. Thus it is not a surprise that the patriarchal tradition of Tampella paper industry management were still embodied in these investments. As far as the content of the investments were concerned, this was indicated by the choice of traditional and technologically undemanding newsprint. According to Ahopelto, the advantage of newsprint was that it used Tampella's long-term experience in this specific paper grade: 'Newsprint is a natural choice for us; we know how to make it and market it and it is likely to have a safe and expanding position in the future.' (*Metsäteollisuusuutiset* 9 September 1980). Another indication of the patriarchal tradition was that Tampella hired only a minimum number of new managers and personnel because of the risk that they would need to be dismissed later.

This patriarchal management tradition reached its limits at the time of the new paper machine start-up in 1983. There were several concurrent reasons for this. First, although Tampella was familiar with newsprint as a product, the technology by which the paper was now produced was significantly more sophisticated than it had been before the new machinery. Mill managers were therefore no longer able to control everything and more and more technical skills and capabilities were required of all personnel. Recruiting trained personnel from outside the mill also became necessary. In addition, the co-operation of personnel was no longer based on loyalty and belonging to the same local community. Instead, the employees and supervisory staff now expected tangible benefits in remuneration for the increased demands of the production process.

Second, the previous management tradition was also inadequate because the lenders and corporate owners expected to profit from the investments which had now finally been made. The knowledge of new technologies became a critical management capability as making efficient use of new production equipment posed significant challenges. Moreover, whereas managers were previously expected only to take responsibility for the manufacturing process they now had to be active in the marketing of the products as well. Unfortunately, this was not noticed before the start-up, because Tampella had relied on Finnpap to market its new capacity. As a result, the lack of prior marketing and sales promotion hampered start-up of the new paper machine (PM 3). Now the experience gained showed concretely that the marketing and product development functions had to be included in the mill management and managers competent in the field had to be recruited.

As a consequence, the Tampella paper industry management started to resemble the management of a modern business. This is indicated by several visible changes in the management of the Tampella mill. In brief, the mill management gained strength and differentiated. For example, the production department was divided into diverse functions and subdivisions. This professionalisation of the production function permitted introduction and development of new incentive systems and troubleshooting groups in the production management. The mill also obtained its first full-term marketing manager and the mill's market procedures were significantly up-dated. Kalevi Aalto was appointed marketing manager at the beginning of 1985 and at the same time his predecessor, Johan Granlund, was chosen to head the mill. As a matter of fact, he was the first person to head the Tampella paper mill without training in paper industry engineering and without management experience in the forest industry. The new market connections of the mill and the new managers recruited to Tampella contributed to the capacity of the mill management to take major responsibility for the business development of the mill.

In summary, before the technological discontinuity in the late 1980s, Tampella's paper industry management was in the process of leaving its patriarchal tradition behind and starting to manage the paper industry as an independent business. The recruitment of new managers permitted the exposure of the mill to more up-to-date ways of thinking in business management. As a consequence, attention turned from the efficiency of production to the efficiency of other functions such as marketing and product development. Thus it also became possible to benchmark the competitors in relation to these other functions and to increase the effectiveness of business strategy and the overall competitivity of the mill in relation to its competitors. As a result, the Tampella mill management would become more capable of taking responsibility for strategic action when the material resources emerged. The roles of the different management levels were now clarified whereas they had previously been confusing. Nevertheless, Tampella still had scant experience of up-to-date paper products and production technology. This was because Tampella had not yet produced them and because most of the recruited managers also lacked such experience. It can therefore be argued that both the managerial capacity and the technological standard of Tampella were still low compared with most of its domestic competitors (see Laurila 1995: 56). Most essentially, Tampella lacked experience in relation to the technological discontinuities it was soon to manage. The details of that discontinuity are provided in the next section.

The content of the discontinuity: introducing new products and production technologies

The section above described Tampella's managerial and technological standing in the paper industry before the late 1980s. This section shows that starting from 1987, Tampella suddenly broke its previous traditions on this field. In brief, we demonstrate here that the technological discontinuity then initiated dramatically contradicted Tampella's past and thus created a clear action gap for the corporation. We begin by describing how the major technological change project emerged during 1987. The latter part of the section explores the ways in which the discontinuity actually challenged Tampella and its management.

The initiation of the technological discontinuity

The formal triggering event for the initiation of the technological discontinuity was the change in the corporate ownership in March 1987. To mention some details, Tampella's main owner until 1987 was the Union Bank of Finland (UBF), which had been the firm's main banker for decades and which gradually obtained a large proportion of Tampella's shares. Now UBF sold its shares to the Skopbank Group which was less established than UBF, but strongly expanding at the time. It had succeeded in its recent take-over operations and was willing to play a new role in financing industrial corporations (see Tainio *et al.* 1991: 198–202). Soon after the take-over Skopbank announced that Tampella was a long-term investment for the bank and that there was capital available for expenditures in different parts of the corporation. This can be expected to have encouraged managers to stay in the company, which had not been quite successful in its recent past. It also motivated them to do further planning and to produce adequate proposals through which the bankers could assess the actual value of their new possession.

The issue for Tampella managers thus was to produce development proposals. The problem, however, was that such proposals cannot be produced quickly. Nevertheless, the idea of extending Tampella's paper industry activities was soon taken up. This was partly because Tampella managers had previously made several calculations of the prospects for improving mill profitability by further investments in production technology. These drafts had been prepared in previous years and aimed to solve the problems that remained after the investments of the early 1980s. In particular, every possible alternative for reducing Tampella's large capacity in newsprint production was examined.

In response to a formal request by the board of Tampella in June 1987, a group of managers set out to draft a plan for the development of the Tampella paper mill. The group reported to the Tampella CEO Jormakka. The parties involved in the process included the managers at the different levels in the formal chain of command (Ahopelto, the head of forest division, and the managers Granlund, Valtonen and Aalto) and a technical consultant. The group also consulted external marketing specialists (managers of the joint sales association Finnpap). The group surveyed the options for altering Tampella's paper product portfolio. Every one of these alternatives had different costs and technical prerequisites. For example, building a completely new paper machine to produce a new paper product requires major extensions to the pulping facilities, whereas replacement of an existing machine with a new one only requires a minor extension. To provide sufficient information on these aspects, the technical consultant prepared a feasibility study which included several alternatives for capital expenditure, for example, calculations for building a new machine and several ways of modifying the existing machinery.

However, the group was to recommend one alternative and it decided on one in which a single paper machine (PM 2) was to be dismantled and replaced with a new one producing MFC magazine paper (machine finished coated), offering substantially higher value than Tampella's existing paper products. Simultaneous modernisation of a second paper machine (PM 3) was also proposed to increase production output and to permit further product development from standard newsprint to slightly more refined newsprint speciality grades (HiFi grades). These changes required major investment in new technology but implied only minor increases in the use of pulp.

This proposal for the development of Tampella's paper industry activities was completed by the end of October 1987 and was soon delivered to the board. Some pressure was aroused by the rumours according to which Tampella's competitors were planning the same kind of modernisations. Eventually, news of the planning leaked to the media in early December (*Helsingin Sanomat* 10 December 1987). A claim which cannot be verified states that one of the Tampella board members was behind the leak with the aim of speeding up the decision-making. Decisions soon followed. First, less than a week later Tampella CEO Jormakka announced acquisition of a larger holding in the pulp producer Sunila; the use of pulp would increase after introduction of the new paper grades. In early January 1988 the board made the formal decision to modernise Tampella's paper mill; the amount involved was FIM 500 million (US$100 million). The decision was made although the board had not had time to learn the full details of the forthcoming changes as it was only one of the many plans Tampella had at that time. Details related to the difficulties of moving Tampella into a completely new paper market area were never presented to the board.

As a result, Tampella had initiated and formulated a technological discontinuity in six months. This was a fast pace for planning investments in the paper industry in general and for Tampella in particular. Planning for Tampella's technological change project in the early 1980s had taken several years. Now we will examine the other ways in which this project deviated from Tampella's previous experience.

The discontinuity compared with previous experience

We will argue here that the discontinuity was challenging for Tampella in at least three broad senses. First, replacement of the old paper machine (PM 2) with a new MFC machine entailed radical technical changes in the production process. The properties of the old and new PM 2 are compared in Table 5.2. The width of the new machine was similar to the previous one, but its planned speed and production capacity were double. Moreover, in addition to being faster the production process with the new machine was also qualitatively different from the earlier one. This was because it included new elements such as soft calendering and coating. The novelty of the new technology was further increased by the fact that these new elements in the paper production process were accomplished in a technologically demanding way. More concretely, the on-line type of paper coating and calendering machine adopted by Tampella is more demanding than the off-line type because malfunctions in the coating part also stop the entire machine. Thus the new machine differed from the type to which management and personnel were accustomed. For example, Tampella had no prior experience of paper coating at that time.

Second, challenges were also produced by the features of the new paper product. The discontinuity implied giving up newsprint production, which had a fifty-year tradition at Tampella. Thus the previous bulk production was to be replaced by the fashionable MFC, which – compared with newsprint (see Figure 5.2) – was a high value-added magazine paper grade launched on the market only a few years earlier. MFC resembled LWC grade (light weight coated), which had become an increasingly popular printing paper grade during the 1970s and 1980s. MFC deviated from LWC only because it was cheaper and less glossy; this was the result of soft calendering and it brought MFC significant market potential. Thus in addition to being much more difficult to produce than newsprint, MFC also had a different market – namely magazine papers. Thus Tampella had to find new customers because it could not expect its previous newsprint customers to switch to MFC.

Third, the discontinuity was also challenging in terms of actually building the new machinery. The fact that the new paper machine was to be located within an

Table 5.2 Comparison of the old and new PM 2

	Property of the machine	
	Old PM 2 (factual)	New PM 2 (objective)
Product area	Machine finished specialities (book paper)	MFC (machine finished coated)
Width	5.46 m	5.4 m
Speed	750 m/min	1,500 m/min
Capacity	85,000 t/a	140,000 t/a
Calendering	Simple	Advanced/On-line
Coating	None	On-line

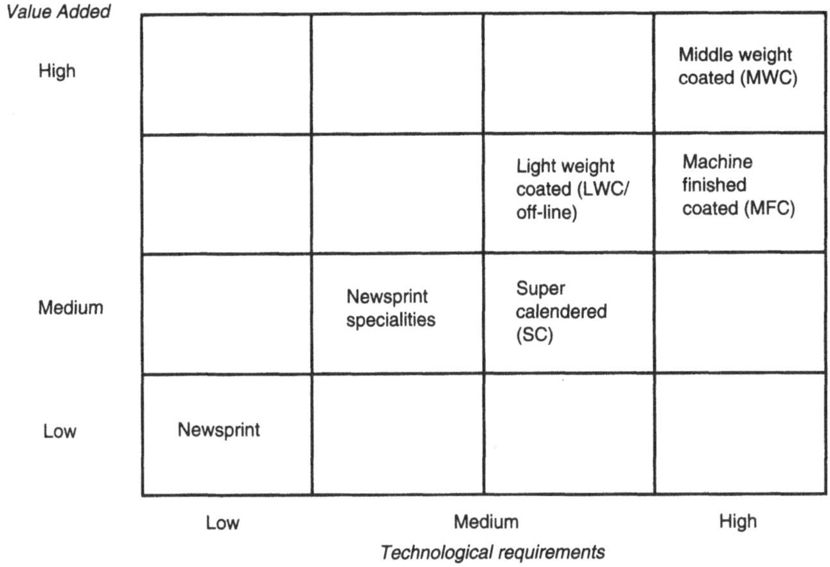

Value Added

High				Middle weight coated (MWC)
			Light weight coated (LWC/ off-line)	Machine finished coated (MFC)
Medium		Newsprint specialities	Super calendered (SC)	
Low	Newsprint			
	Low	Medium		High

Technological requirements

Figure 5.2 Wood-containing paper grades in relation to value added and the requirements of production
Source: Adopted from Laurila (1997a: 227), reprinted by permission of Blackwell Publishers Ltd.

existing facility shortened the time needed to make the changes. Most often (when new pulping capacity is not needed), however, the replacement is made by building the new paper machine alongside the old one and letting the old machine run until the start-up of the new. Because this was not the case here some potential production and cash flow was wasted. In fact, the interruptions in cash flow would double because the costs of building the new machine had to be covered with declining revenues. The cash-flow effect of the Tampella PM 2 investment is illustrated in Figure 5.3. Thus the changes necessitated extensive financing though Tampella was already heavily indebted even without new capital outlays (see Table 4.2 in the previous chapter). Moreover, some losses were also caused by the fact that the old paper machine (PM 2), which was now to be dismantled and sold abroad for scrap, was not technically exhausted. Instead, it was slightly more advanced than the similar PM 1 and was producing Tampella's most profitable paper product, book paper. As a matter of fact, PM 2 was chosen instead of PM 1 because otherwise the necessary physical connection between PM 2 and PM 3 would have been blocked. Finally, the new paper machine had to be built at a running mill, and this would obviously hamper normal operation of the other two paper machines and also the start-up of the new unit.

As a matter of fact, the potential problems posed by the initiated technological discontinuity gave rise to a competing idea of how to move Tampella into coated paper grades. This would have meant acquisition of only one or two

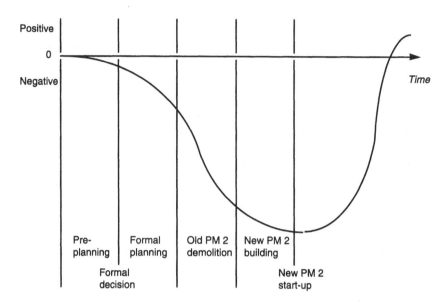

Figure 5.3 Cash flow effect of the Tampella PM 2 investment

off-line coating units instead of an entire paper machine. In other words, the existing products would have been refined instead of introducing a totally new paper grade. This idea had been proposed by production manager Valtonen, but was not supported by the other managers, even though they were aware of the many difficulties experienced by the other Finnish mills in starting MFC production a short time earlier.

To summarise, the initiated technological discontinuity in many ways contradicted Tampella's previous experience and implied many challenges and hazards. The basic motivation for the discontinuity was the acquisition of new technology that would permit the mill to move into significantly more refined and more highly priced products at one time. If successfully implemented, the concept would be a cost-efficient way to produce coated paper partly because of the small number of personnel needed in the production process. Entering an unfamiliar market could also be expected to be less difficult because Tampella used a product concept similar to that of the other Finnish MFC producers.[2] The fact that customers want alternative suppliers would make Tampella's market entry easier. It could also be argued that not all Tampella paper machines would be rebuilt and also that the changes on PM 3 would not start until after the changes on PM 2 were completed.

Nevertheless, it is obvious that the discontinuity comprised significant technological and market risks and also a technically difficult building process. Both the managers and the personnel of the Tampella mill were unfamiliar with this new technology and the markets in which the new products were to be sold. Moreover, the technological sophistication of the MFC grade also made it technically inconvenient. As stated above, the process included several new elements, some of which had never been used at the Tampella paper mill before. Now these new elements were to be combined in a continuous high-speed production process. Moreover, as the last section of this chapter will demonstrate, this was not enough. The management of Tampella had the courage to be even more ambitious and increase the risks inherent in implementing the technological discontinuity. Duplicating its previous achievements was not sufficient for Tampella at that stage; it decided instead to make the changes even more challenging by building the machinery at its own engineering works. This decision was made despite the fact that at that time Tampella had made neither printing paper machines nor machinery for coated paper grades.

After specifying the challenge offered by the technological discontinuity, the study now proceeds to examine the implementation of this discontinuity and how it was coupled with changes that occurred on the different levels of managerial hierarchy.

The implementation of the discontinuity

Implementation of the technological discontinuity in focus has two distinctive features. First, in 1988–1990 it faced severe delays and difficulties which seem to be related to the quick decision to adopt new technology, the tight implementation schedule and the overall ambitiousness of the discontinuity, which was further increased during implementation. Second, during the period between 1988 and 1990 virtually all managers who had originally taken part in the initiation and preliminary planning left or were transferred to other positions within Tampella. This section briefly analyses these features. It starts with the details of the new technology adoption and closes with an examination of parallel changes in the management composition.

Difficulties in new technology adoption

In terms of production technology, there were only preliminary technical plans when implementation of the technological discontinuity began in early 1988. These plans were gradually specified. At the outset, the product concept and market segmentation were not clearly defined and the project did not have a detailed budget. The detailed planning, however, therefore produced many surprises which transformed the original idea. Since none of the existing parts would fit the new design the entire machine had to be replaced. This was partly because the planning process revealed that the speed of the new machine could be increased from 1,200 metres per minute to 1,500 metres. As a result, production could be increased from the original 110,000 tons to 140,000 tons per year. At this level of production the Tampella PM 2 would have had the world's largest output in its width category.

Although some problems were expected, the number of technical difficulties involved in building the new machine within an existing facility was another surprise. For example, the physical proximity of the existing paper machines (PM 1 and PM 2) made the replacement of the latter more difficult. The difficulties were multiplied by the fact that during the project there was a boom on the paper market and all of the machines (PM 1, PM 2 and PM 3) were running at full capacity. It was therefore difficult to arrange the stoppages necessary for planning and installation.

The planned start-up of the new PM 2 was already postponed in the early phases of the project from the original objective of summer 1989 to the end of October 1989. In addition, because of the increase in the planned production capacity of the new PM 2, its overall budget combined with the PM 3 renovation was extended from 500 to roughly 600 million FIM. In principle, Tampella was free to choose any technology supplier. However, termination of the earlier co-operation between Tampella and the other two Finnish paper machine manufacturers – Valmet and Wärtsilä – (see Chapter 7 of this volume) significantly restricted this freedom in practice. First, as Tampella's Pulp and Paper Machinery Division had entered the paper machine market in 1987 (it had previously delivered only board machines) it badly needed paper machine references. Second, relations between Tampella and Valmet in 1988 were so bad that Valmet (which had supplied Tampella PM 3) did not even bother to make a tender for PM 2.

As a consequence, at the end of the summer 1988 it was decided that the new PM 2 would be delivered by Tampella's own engineering works. In fact, this was going to be only the second printing paper machine ever built by Tampella. This also meant a further increase in the risks inherent in the technological discontinuity. In other words, both the receiving and the supplying halves of the project were unfamiliar with the new technology. The fact that the new technology was delivered to the Tampella paper mill by another unit of the Tampella Corporation had several consequences. On the one hand, because Tampella's Paper Machinery Division was able to supply only the core machine, many parts of the order had to be subcontracted. Thus co-ordination between many different subcontractors became more complicated. Moreover, because of the inexperience of the main supplier there were no tested solutions to the many problems related to the new technology. On the other hand, competition between the remaining suppliers decreased because the Finnish suppliers did not make an offer. This was understandable because they did not want to help their competitor Tampella to enter the paper machine market by submitting technical information on their own machinery. Consequently, the service offered to Tampella by the subcontractors was probably inferior and the price offered higher than in a situation in which domestic suppliers would also have been competing for the same order.

These various unfortunate details started to have a visible impact from early 1989 on when Tampella's paper production efficiency started to decrease. As a matter of fact, a record volume of production was achieved at the Tampella paper mill during the previous year: 380,000 tons of paper. However, installation of a paper machine in a mill running at full capacity was part of the reason for the decline in output. Later that year it was clear that the start-up of the new PM 2 would again be delayed, this time by about three weeks. This was because Tampella's Pulp and Paper Machine Division was also supplying another paper machine which was to start almost at the same time to one of its domestic paper industry competitors. Thus the manufacturing capacity of the Tampella engineering works was temporarily exceeded and it was unable to deliver all the machinery to the Tampella paper mill as negotiated.

In the end, start-up of the new PM 2 was delayed by six weeks. This is to say the first roll of (still uncoated) paper was produced after a month of uninterrupted effort. Some of the difficulties were related to the high running speed at which the machine started. The first rolls of coated paper were produced in January 1990. However, the quality of the new product did not allow deliveries of samples to the customers for another two months. This delayed entry to the market. Moreover, because PM 2 had to be repaired after start-up, the project exceeded its overall budget by about FIM 50 million (US$10 million). Technical malfunctions continued throughout 1990. Production on PM 3, which was restarted in summer 1990 after less profound alterations, was also plagued by technical problems.

The short-term effects of the technological discontinuity were disastrous. Apart from the failure of start-up to meet expectations, disruptions in production reduced Tampella's overall paper output to a low level for 1989 and 1990. In these years the paper output was at least 100,000 tons less than expected. The resulting losses were approximately 150 million FIM (US$30 million). The difficulties also continued after this. The new PM 2 produced only 200,000 tons of MFC in three years (1990–1992), compared with a theoretical production capacity more than 400,000 tonnes. In addition, in 1994, after over four years of operation, the new PM 2 was still 50,000 tons behind its theoretical annual production capacity. At the same time its production efficiency was some 25 per cent below that of its closest competitors, which had started their machines a couple of years earlier.

In summary, implementation of the technological discontinuity at Tampella was difficult. Although the building and installation of the new technology succeeded in principle, the paper output fell short of the ambitious objectives, at least in the short term. From the perspective of this book, however, it is essential to give an overview of what was happening at the same time at the different levels of Tampella's managerial hierarchy. After describing the various technical difficulties in the implementation and the apparent discrepancies between short-term performance and expectations, this section therefore moves forward to examine the way these features were coupled with changes in management characters.

Changes in management characters at different management levels

During the implementation of the technological discontinuity in 1988–1990 there were numerous changes in the management characters at the different levels of Tampella's managerial hierarchy. In general, this was because in early 1988, after the formal decision to adopt new technology, Tampella had the financial resources to start the actual project, but it had only limited managerial resources for the same purpose. There were at least two broad reasons for the changes. First, changes were due to the need to hire new managers with the capabilities that Tampella was now lacking. In other words, adoption of new technology created new managerial positions which could not be filled by the existing staff. For example, three experienced engineers were hired for the marketing and product development function at Tampella paper mill and one civil engineer with some previous experience of MFC production was appointed to work in the building of the new PM 2. Moreover, during 1988–1989 several new managers were also hired for various business development activities at different levels.

The technical head of the project, however, was obtained from within the company. Risto Jouppila (an engineer with extensive experience at the mill and wide support among the mill personnel but without any experience in paper coating technology) was chosen to head the machine building project. This meant that the management of the actual project was delegated to the best men available, picked from among the most experienced production engineers at the mill. There were also many open positions in the early stages, because the project had no full-time managers. In fact, no new managers were recruited before the formal decision and therefore the project was initially managed along with normal routines. For example, general management of the project was the responsibility of the mill head, Granlund, and marketing was the responsibility of Aalto.

Second, in addition to the new positions replacements of previous managers necessitated extensive recruitment. The first of these replacements occurred only two weeks after the formal decision on the technological discontinuity in focus here. This was the replacement of the previous Tampella CEO Jormakka with Paavo Lonka. In contrast to Jormakka and other individuals involved in the initiation of the technological discontinuity, Lonka was experienced in numerous technological change projects as he had worked in different positions within the management of the Finnish paper machinery industry (at Wärtsilä and Valmet). This replacement significantly increased the number of new managers at Tampella as Lonka recruited new people who then continued this process (see Laurila 1992: 72). According to Lonka's proposal, Tampella's board decided in the summer of 1988 to replace the long-term head of the forest industry division Ahopelto with Aulis Lehtonen, who had extensive experience of managing technological change projects at other Finnish paper mills.

The new head of Tampella Forest Industry Division gave his full support to the major technological change project already started at the Tampella paper mill. He also supported the above mentioned modifications in the project which permitted significant increases in the planned paper output. However, he was clearly disappointed with the technical problems that soon emerged and caused financial losses. In late 1989, when both Tampella's paper production efficiency and the actual construction of the new facilities were lagging behind schedule, he decided to replace Granlund, the head of the Tampella paper mill with Sven Nylund. The latter was a previous colleague of Lehtonen from the time when he was employed by another Finnish paper industry company. Nylund, who was not an engineer and who did not have experience in the management of paper industry investment projects, was then faced with the responsibility of managing a mill which had raised great expectations, but which faced immense difficulties. The reorganisations he initiated in early 1990 included demoting the holders of previous key positions. For example, technical project manager Jouppila was appointed purchasing manager and production manager Valtonen was appointed head of maintenance operations.

In early 1990, the new head of Tampella paper mill initiated division of the mill into two business units. This reorganisation created new managerial positions which were occupied mostly by recently recruited managers. For example, the previous responsibilities of production manager Valtonen were divided between two significantly younger and less experienced engineers. As a matter of fact, the new managerial positions at the mill were occupied by managers whose average age was slightly over thirty. In addition, there were several managers whose experience was limited to a couple of years of employment with some of Tampella's domestic competitors; half of those appointed had been at Tampella for less than two years. These reorganisations continued the process in which the managers who had initiated the technological discontinuity were replaced one by one. Four of the five managers who were involved in the preparation of the investment proposal in the autumn of 1987 were replaced or transferred to other positions before spring 1990 (see Figure 5.4).

In summary, Tampella managers were highly mobile during the implementation of the technological discontinuity. New managers were coming in to either replace previous Tampella heads or to supplement their lacking capabilities. The replaced managers either stayed in their new staff positions or left the company. Most essentially, however, the mobility and replacements of managers indicate the managerial challenges inherent in a technological discontinuity.

Conclusions

This chapter provided an overview of the background, initiation and implementation of the technological discontinuity in focus. The main events of Tampella's paper industry development are presented in Table 5.3. How should these details then be interpreted? The material examined here has brought out a few essential features. First, from the technology point of view, the technological discontinuity at Tampella can be considered a laggard's late attempt to close the gap with its competitors and thus to restore the mill to the front rank of paper producers. Because of Tampella's inexperience with leading edge technology in paper making, the change project – which was not unusual for the Finnish paper industry (cf. Rohweder 1993; Laurila 1992: 45) – can be considered extremely ambitious. Despite the fact that the discontinuity failed to reach its

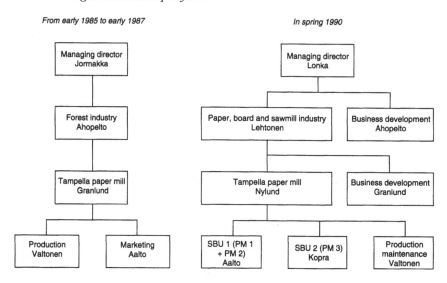

Figure 5.4 Tampella's formal chain of command with reference to the Tampella paper mill

original objectives it should be considered a major success in terms of the mill's previous experience. In fact, in a short time Tampella acquired several capabilities it had previously lacked and as a consequence almost reached its closest competitors, which had developed their production from much better starting points. They either had several paper mills between which knowledge and managers could be transferred or used suppliers that had already provided the new technology in question to other customers.

Second, and more importantly, the details presented demonstrate that the technological discontinuity at Tampella was essentially a managerial challenge. It should be noted that the individual managers at Tampella supported the discontinuity despite the apparent risks it implied for their future careers as their relevant experience was limited. In a sense, these managers took the risk of being replaced in the event of difficulties or simply by not being an expert on the new field of paper production. However, due to the high level of mobility and the fact that several individuals and groups of individuals were involved in the process, the technological discontinuity at Tampella is an opportunity to solve the research problem of this study; namely to explore how managerial actors from different backgrounds came to promote the discontinuity, became motivated and committed to work for the venture and gained the courage to increase the risks involved to an extremely high level. To add such a mobilisation perspective of this kind to the discontinuity at Tampella, the interests of the managerial actors involved must be incorporated into the analysis. The next chapter will therefore analyse the managerial actors and their mobilisation in the setting described in this chapter.

Table 5.3 Main events of Tampella's paper industry development 1980–1991

21 May 1980	The formal decision on building PM 3
1 January 1983	Mr Jormakka, the new CEO, takes over after the planned retirement of his predecessor Mr Gadd
25 February 1983	Start-up of PM 3
February 1987	Tampella decides to move into production of paper machines
12 March 1987	Announcement of Tampella's take-over by Skopbank
24 April 1987	Appointment of a new board of directors for Tampella
7 January 1988	The formal decision to rebuild PM 2 by Tampella's board
1 February 1988	Mr Lonka, the new CEO takes over after his predecessor Mr Jormakka had been dismissed on 14 January
June 1988	Launching of the corporate turnaround programme
December 1988	Tampella's profit doubles compared with the preceding years
5 April 1989	Announcement of the improved corporate profitability for 1988
9 August 1989	The board discovers that corporate profitability is declining
November 1989	Announcement of the first profit warning for Tampella
17 December 1989	Start-up of the new PM 2
February 1990	Launching of the 'From Talk to Action' programme
June 1990	All Tampella's ranges of business have difficulties with their performance
August 1990	Severe problems in corporate profitability
20 September 1990	The board decides to replace the corporate CEO
14 December 1990	Formal dismissal of the corporate CEO
8 January 1991	Mr Sorri takes over as the new corporate CEO
21 January 1991	All Tampella's ranges of business are in crisis
4 February 1991	The Finnish government no longer allows Skopbank to finance Tampella
27 February 1991	Public announcement of the decline in corporate profitability for 1990
9 April 1991	Three outside members of Tampella's board resign
19 September 1991	Take-over of Tampella and Skopbank by the Finnish government

6 The actor perspective to the discontinuity

How did management become mobilised?

This chapter analyses managerial actors and their mobilisation to initiate and implement the technological discontinuity described previously. It therefore takes a managerial actor perspective on the discontinuity. The managerial actor perspective refers to the subjective experiences and motives of the managerial actors involved. The chapter comprises four sections, each of which has a specific objective in the conceptualisation of the mobilisation process.

The first section describes the type and characteristics of managerial actors involved in the discontinuity. It aims to identify the background of each actor and the managerial objectives promoted by their actions. The second section of this chapter moves from a characterisation of the actors to the characterisation of their relations to the specific technological discontinuity at hand. In other words, it conceptualises the mobilisation of the various actors for the discontinuity at Tampella by examining how the discontinuity was congruent with the managerial objectives of the actors, thereby permitting them to promote these objectives.

The third section of this chapter develops the analysis of managerial mobilisation further by analysing how the constellation of actors initiating the discontinuity becomes characterised by an accelerating level of ambition and managerial courage. Thus the section brings in the socio-emotional factor of mobilisation by identifying the personal values and goals of the actors. As they are aligned with the managerial objectives, they bring about psychological commitment which leads to temporary energisation within management. This dynamic phenomenon will be analysed both in the increasing managerial effort involved in the discontinuity at Tampella and then in the subsequent disintegration of this temporary coalition. The fourth section of this chapter elaborates on the analysis and comments on the mobilisation of managerial actors in general. It focuses on the nature of the mobilisation at Tampella as a coalition of several micro-mobilisations and the factors which functioned as triggers for mobilisation and obstacles to its continuity.

Characterisation of the managerial actors involved in the discontinuity

The characterisation of the managerial actors provided in this section is based on the ideal type construction method (Weber 1947) which exaggerates the features of individual managers to produce a distinctive quality for each collective actor. This is to say the actors described are aggregates of individual managers, some of whom were mentioned in the preceding chapter. The actors thus have their reference points in the individual managers. The sayings and doings of these individual managers are also used to illustrate the nature of these actors. However, the boundaries between different actors are indefinite in the sense that individual managers may have the characteristics of different actors. We therefore also describe the characteristics of a typical manager for each actor. This typical manager is not, however, any of the actual individuals.

The managerial actors involved in the technological discontinuity in focus will be examined in the order of their actual appearance. Consequently, the analysis is once again presented in three subsections. The first subsection analyses the constellation of actors before the changes in Tampella's ownership in early 1987. The second subsection presents the new owners and how their emergence changed the relationships between the original actors in the initiation phase of the discontinuity. The third subsection introduces two actors who entered during the implementation of the discontinuity between 1988 and 1990.

The constellation of managerial actors before the discontinuity

A dominant actor leaving the stage: the irresolute owners

Before early 1987, the irresolute owners, namely the Union Bank of Finland (UBF) and its associates, the key owners and financial backers of Tampella, were the dominant actors in the management of the company. Although they were not directly involved in the discontinuity in focus, they must be mentioned; otherwise the characteristics of the other actors cannot be understood. More concretely, the owners influenced the formation of the other actors by appointing Tampella key managers and by deciding on the financial resources granted for each corporate activity. These decisions greatly affected the level of cumulated technological and managerial capabilities at the time of initiation of the discontinuity in focus.[1]

The irresolute owners were highly sceptical towards Tampella. They had a long tradition of helping Tampella through difficult periods and financial setbacks. They knew that Tampella operated in highly cyclical industries and that every major technological change project was a financial risk. They were also aware that Tampella was not one of the technologically most advanced Finnish paper industry companies. UBF hired technical consultants to examine Tampella a short time before the change in company ownership. One of the managers describes the outcome: 'Pöyry [the technical consultancy firm] said to us: "My God, sell this stuff!" ' (Saari 1992: 117). The owners were therefore ready to give up Tampella. However, even before the ownership change they were not totally unhappy with the situation; although Tampella Corporation was in debt it was still meeting its liabilities and paying dividends. In other words, Tampella provided a steady return to its owners, who were in general not motivated to undertake strategic endeavours requiring significant capital expenditures.

The owners had two main ways of controlling the risks inherent in Tampella. Apart from the fact that the representatives of the owners used their power on Tampella's board of directors to reject risky and ambitious investments, they also appointed key executives who they thought should be able to restrict risks. For example, Tampella's CEOs in the 1970s and 1980s clearly preferred to avoid radical changes in Tampella's corporate structure and instead to undertake international ventures focused on detaching the corporation from its earlier unsuccessful ones. Tampella CEOs during this period were quite different from the entrepreneurs who had expanded the corporation in the previous decades.

The unwillingness to increase the owners' liabilities in Tampella was one reason why the strategic problems of the corporation in general and those of its paper mill in particular could not be solved. As the previous chapter demonstrated, Tampella was technologically behind its domestic paper industry competitors before the technological discontinuity. However, from the irresolute owners' perspective, these problems were minor compared with those faced by the company as a whole and in addition, the mill's cash-flow was positive. There were two other reasons why the development of Tampella's paper industry operations was not the main interest of the irresolute owners. First, the paper mill was the most capital-intensive part of the corporation and keeping it competitive would have required significant capital expenditures on a recurrent basis. Second, the building of the new facilities at the mill in the early 1980s, which saved the mill from total collapse and which had been promising for the company as a whole, had not met the owners' expectations. However, because the irresolute owners financed this project mainly by borrowing, their losses from the investment were minor. This is not to say that the owners would not have contributed to these shortcomings themselves. This is because they also controlled other Finnish paper industry companies whose interests would be protected by preventing Tampella from investing in the more value-added printing paper grades like LWC. These grades were, in the unofficial strategic division within the Finnish paper industry, reserved for other companies.

The reason for analysing the characteristics of the irresolute owners is that they are related to the nature of the actors responsible for the management of Tampella at the time of the ownership change. The actions of Tampella managers had to be at least partly congruent with the conditions created by the existing ownership. In other words, the actors somehow adapted to the circumstances before the discontinuity in focus. At that time it is possible to identify three managerial actors with distinctive means and objectives in the management of Tampella paper mill. These actors can be labelled in the order of their emergence as technology enthusiasts, production functionalists and business innovators. The characteristics and managerial objectives of these actors will be examined in more detail below.

Technology enthusiasts: keeping the faith on technological modernity

The technology enthusiasts were paper industry engineers located on the different hierarchical levels of Tampella. Most often they had been employed by Tampella's various forest industry units for their entire careers. Some of the individual managers had also lived through Tampella's hard times of technological decline during the 1960s and 1970s. They had remained with the company, although they might have found more interesting jobs elsewhere. For some of the technology enthusiasts this was no longer a realistic option because although their technical expertise had originally been of the highest degree, their knowledge had gradually become outdated because there had been few investments in paper production technology at Tampella. Moving elsewhere would also have meant losing their visible position at the Tampella community. Consequently, in the spirit of Gouldner (1957), technology enthusiasts were clearly locals.

As far as the general managerial objectives were concerned the technology enthusiasts were characterised by the belief that the critical factor for success in the paper industry is the modernity of the production technology and paper machinery. This is because the largest economics of scale are gained with the fastest and widest paper machines on the market. The individual technology enthusiasts had seen how the competitors with their more advanced machines had gradually pushed Tampella into a more and more difficult position. They had noticed that technological development will always bypass still serviceable, but obsolete machinery. The contradiction between these lines of thinking and the fact that Tampella no longer represented the state of the art in paper making is reflected in the following comment by an individual manager: 'As [Tampella's] first paper machines were acquired in 1937 and in 1938 it is odd that the next one was acquired in 1983. There should have been one or two machines in between' (Marttila 1988: 107).

How then had the technology enthusiasts rationalised this contradiction between belief in technological modernity and Tampella's relative technological backwardness. The fact that Tampella had always been the last in the Finnish paper industry to invest did not discourage them because they believed that the latecomer gets the most advanced technology. As one manager put it: 'If we [Tampella] had acted according to the original [investment] plans [in the late 1960s], we would not have the good equipment we have now' (Marttila 1988: 106–107). Technology enthusiasts thus considered that cancellation of investment proposals was in fact only a postponement. Another factor to reduce the technology enthusiasts' anxiety was that they were little concerned with what was produced with the modern production technology. Thus the minor improvements in production technology and especially the new newsprint machine in the early 1980s could be considered progress. The technology enthusiasts did not think it was necessary to alter the product portfolio of a paper mill to sustain competitiveness. From this perspective, even the production of low value-added paper grades would be appropriate if the production technology was competitive. The technology enthusiasts considered Tampella one of the best newsprint producers. Consequently, from the technological enthusiasts' point of view, the key aim was to keep Tampella technologically modern through recurrent machinery rebuilds.

The fact that Tampella did not perform well after the start-up of the new machinery in the early 1980s did not threaten the belief of the technology enthusiasts in technological modernity. The rationalisations protecting their belief included the fact that the long payback time for investments in paper industry technology makes it difficult to predict their profitability; prices for the final products and the costs of production may vary significantly in time. As a result, the technology enthusiasts believed that the overextension which took place on the newsprint market in the early 1980s had punished competitors as much as Tampella and that the technological leader would always prevail in the long run. Moreover, during their careers some of the technology enthusiasts had co-operated with Tampella's Machinery Division, which had usually supplied most of the technology used at the forest industry facilities of the company (see this volume, Chapter 7). This co-operation provided the technology enthusiasts with a steady stream of new ideas for machinery.

The production functionalists: emphasising the production process

The production functionalists were technically educated managers with long-term management experience at Tampella. Most often they had not reached the highest ranks in Tampella's hierarchy. This was because they lacked either the necessary education or sufficiently long working experience. Their educational background was in fact relatively weak compared with the general standards in the Finnish paper industry; most of them lacked higher degrees in paper industry engineering. However, they had been able to move up the managerial ladder because not many managers in their field were hired by Tampella before the technological discontinuity in focus. This was also possible because their responsibilities had been somewhat adjusted to their capabilities. Because of their long technical experience in managing the concrete production process, they had strong local relationships and some of them were also well-liked at the shopfloor level.

Whereas the technology enthusiasts emphasised the modernity of the production technology, the production functionalists emphasised the fluency of the production process. Their strategy for surviving in Tampella was not to wait for major change projects and investments, but instead to enjoy all progress that they could make in improving the efficiency of the existing production facilities. According to the dominant way of thinking among the production functionalists, incremental improvements in production efficiency were essential to the long-term profitability of a paper producing company. Experience had taught them that efficient use of new technology demands hard work. The production functionalists accomplished their aim through *ad hoc* trouble shooting and knowledge exchange between Tampella and its competitors.[2] In this work they had experienced that through innovations in the production process it is possible to improve the production efficiency of a paper mill without noteworthy capital expenditures. For example, significant improvement can be made simply by reducing the number of paper grades in production and by producing technically similar grades together.

The business innovators: focusing attention on products and customers

Whereas both technology enthusiasts and production functionalists emphasised categories related to production technology, the business innovators emphasised the paper products and the customers to whom they were delivered. In contrast to the other actors, the business innovators had only limited experience in the management of Tampella in general and of its paper industry activities in particular. Before entering the paper industry most of them had worked in other businesses inside and outside Tampella. Consequently, they did not have strong local connections at the Tampella paper mill. Moreover, most of the individual business innovators lacked previous training or experience in the paper industry. They were therefore not specialists in paper industry technology or even paper industry products. Nevertheless, with the help of their general management and marketing skills they could make a significant contribution to Tampella's paper industry management. Their general skills enabled for example development of incentive schemes and other management procedures. Their interest in contacts with paper consumers was also important as the marketing and sales functions were simultaneously being transferred from the joint sales association Finnpap to the Finnish paper producing companies.

The major belief of the business innovators was that adjusting a paper mill's products and service to better meet the expectations of customers was an important, though previously neglected way to improve performance. This belief is reflected in the following statement by an individual manager:

> There are two ways to react to a customer's wish. The first is to argue why this or that cannot be done. Another way is, whenever possible, to try to accomplish the customer's wishes. . . . I think the latter is the only way to improve profitability.
>
> (Marttila 1988: 156)

The business innovators' experience in other business settings helped them to recognise the lack of business management in the Finnish paper industry in general and at Tampella in particular. More specifically, they were able to criticise the existing products and management methods at the Tampella paper mill. They argued that in Finland low value-added papers such as newsprint would not – with any technology – secure profitable paper production in the future. Obviously, the business innovators would have faced the same problem as the technology enthusiasts, as such changes in the paper industry setting usually require significant outlays of capital.

To summarise, we have now identified four actors in the management of Tampella before the technological discontinuity in focus. One of them, the irresolute owners, were not directly responsible for the management of the Tampella paper mill and in addition, they were soon replaced by the new owners. The other three actors differ from each other for example in respect of their underlying beliefs, focus of attention and the nature of their previous experience. The most important implication of these divisions for the theme of this book is that the actors originally disagreed on the principles concerning the further development of Tampella's paper industry activities. More specifically, the actors were divided in relation to three main issues which are presented in Table 6.1.

First, the object of change separated the business innovators from the others because they focused on products whereas the technology enthusiasts and production functionalists focused on production technology. However, the technology enthusiasts and the production functionalists were also divided in the sense that the former emphasised the essence of major renovations in the production technology. Thus they were less interested in the actual production process, and mainly compared investment in different printing paper grades without reference to the capabilities needed for their production. On the contrary, the production functionalists emphasised the incremental process development that is the basis of increases in production efficiency. Second, in the view of the technology enthusiasts, the optimal type of change was discontinuous in the sense that modernisation should only take place at long intervals. More concretely, they believed that through major investments the production technology of a paper mill would remain competitive for a long time. In contrast, both production functionalists and business innovators agreed that products and production equipment must in any case go through minor changes at frequent intervals.

Table 6.1 The traditional sources of conflicts in the management of the Tampella paper mill

Issue	Actor		
	Technology enthusiasts	*Production functionalists*	*Business innovators*
The object of change	Production equipment	Production process	Service to customers
The nature of change	Periodical	Continuous	Continuous
The basis of change	Technological intra-firm linkages	Learning by doing	External feedback and imitation

Third, the basis of development can either be internal or external. In this respect, both technology enthusiasts and production functionalists emphasised the undisturbed operation of the mill instead of the acts of competitors and the requirements of markets. For production functionalists, the diverse product range threatened the efficiency of the production process. In contrast, for business innovators it meant the creation of long-term relationships with customers. In other words, production functionalists were more concerned with difficulties in adopting new paper grades and their production processes and the business innovators with meeting customer expectations. There is still another related difference. We noted above that the business innovators promoted the emerging business management of Tampella's paper industry. To advance this process, they were willing to recruit new managerial resources for the mill. In contrast, the technology enthusiasts in particular had been reluctant to hire new managers. However, both because of their own previous experience and the recruitment of new managers the business innovators opened Tampella to outside impulses, whereas the other actors had a more internal orientation.

Thus there were conflicts between the actors involved even though only a few technological changes were actually effected during the period before the technological discontinuity. Moreover, the relationships between these actors would change soon after the new owners took over Tampella in 1987. The characteristics of these risk-seeking owners are therefore examined next.

The constellation of managerial actors in the initiation phase of the discontinuity

Creation of a new dominant actor: the risk-seeking owners

The new owners of Tampella, namely Skopbank and its associates, are here called the risk-seeking owners. This is because in contrast to their predecessors, the risk-seeking owners systematically developed Tampella through a variety of investments. The individual risk-seeking owners were bankers educated in commerce or law. Many of them had long experience in their field, although in fact most of them were rather young for such high formal positions. The risk-seeking owners were experts in finance without local or other bonds to Tampella.[3] The lack of experience in the management of industrial companies did not prevent the risk-seeking owners from seeing that Tampella was not in good shape, at least at the time of the take-over. Although the new owners did not have extensive reports on Tampella before the acquisition, they knew that UBF had exploited the corporation by not supplying major resources and by charging high interest on Tampella's loans. Nevertheless, Tampella's profitability was increasing and the outlook for the future was encouraging.

The take-over of Tampella by the new owners did not take place overnight. In fact, they originally (between 1985 and 1987) had only a small proportion of Tampella's shares, although this proportion grew steadily. When the UBF managers noticed that Skopbank was interested in Tampella, they soon offered their shares to Skopbank. The irresolute owners' willingness to relinquish their holdings in Tampella had also increased gradually. In the structural reorganisations which took place in the Finnish economy in the middle of the 1980s, there was considerable speculation concerning mergers between the divisions of Tampella and other Finnish forest industry corporations. However, no arrangement that would have been suitable for Tampella and its irresolute owners was found in the middle of the 1980s. This was partly due to the fact that Tampella's articles of association prevented the dominant owner from exercising all its power in the company.

The importance of the risk-seeking owners for the further development of Tampella and its management was based on the fact that they differed from the other actors involved and they now had the will and formal power to change Tampella. For the new owners, the take-over was part of the general aim of building a bank sphere of influence with a 'flag ship' company, keeping in with the practice of some Japanese conglomerates and also the other Finnish commercial banks (see Chapter 7 of this volume). However, the take-over of Tampella was also based on the belief that a major increase in the value of the corporation would be possible after reorganisation. In other words, the new owners began to believe that if other plans failed, they would always have the opportunity to make a profit by selling the company as a whole or in several parts. Thus they sought to make Tampella a solid company, which it was not at the time of acquisition.

The new owners became accustomed to the intensive pace of action in modern finance and in their previous operations they had seen that major profits or losses can be made overnight. They soon found out after taking over Tampella that major industrial corporations operate differently. Moreover, they had only a limited number of ideas about what to do with the company. Most of the individual bankers' related experience was from financing small and medium-sized industrial firms. In addition, though they had previously succeeded in making profits with increases in the value of the companies whose shares they had acquired, they had not yet interfered in the management of those companies. In fact, the risk-seeking owners lacked knowledge of the management of an industrial corporation in general and the specific businesses of Tampella in particular. In consequence, there was no clear vision of how to make Tampella profitable in a short span of time. In brief, the new owners promoted business development based on increasing financial assets, but in the conversion of this objective to concrete acts they were dependent on the knowledge and action of the acting management of the corporation and the outside experts they had recruited to Tampella's board of directors. The dependencies of the new owners on the acting management is reflected by the fact that despite having formal power during their dominant period between 1987 and 1991, the new owners did not initiate changes in the main business structure of the company.

Although the new owners did not initiate substantial changes in Tampella's paper industry activities they significantly influenced the company by changing the constellation of acting management. They did this in two ways. First, they changed the relationships between the three actors whose characteristics were analysed above. The actors had previously more or less followed the formal hierarchy, that is, technology enthusiasts conformed to the irresolute owners' aspirations and the production functionalists and business innovators continued their incremental development work. In contrast, the business innovators (who did not have the highest formal power but who had the most creative ideas for the further development of Tampella's paper industry activities) became the most influential of these three actors. The other actors were also gradually persuaded once the support of the new owners for the ideas presented by the business innovators became obvious. Second, the new owners also changed the constellation of managerial actors by initiating recruitment of new managers to Tampella. The new owners recognised that the pace of change they wished to see at Tampella was not in line with the capabilities of the existing management. In brief, from the new owners' perspective Tampella had too few competent managers for the future they encouraged for the company. In particular, after the decision on the technological discontinuity at Tampella paper mill, the company needed new managerial capacity. The next subsection examines the actors who joined the management of Tampella during the implementation phase of the technological discontinuity between 1988 and 1990.

The constellation of managerial actors in the implementation phase of the discontinuity

Features of the newcomers: the experienced generalists and the technical specialists

The new managers who joined the different management levels of Tampella after the ownership change were of two kinds. We refer to the first kind as experienced generalists. Most of them were technically highly qualified and had more or less extensive experience of state-of-the-art technology in the service of some of Tampella's competitors in the Finnish forest industry sector. Nevertheless, they were younger than the typical technology enthusiasts described above. The experienced generalists also extended this technology-focused knowledge by general management experience. This allowed them to take a more strategic view of business development. Typically, they had not been afraid of changing jobs although traditionally engineers in this field served one employer for their entire career. The experienced generalists were seeking challenges such as the management of major technological change projects. Consequently, they had had a successful career through mobility between top hierarchical positions in different paper industry companies or associations. They did not have previous bonds to Tampella, but they were well known within the Finnish paper industry engineers' community.

The experienced generalists promoted the dominant recipe of the Finnish paper industry at that time; namely introducing more value-added paper grades and increasing market and product segmentation through application of new production technology. Because of their experience these generalists were familiar with the processes of designing new generation machines and their implementation in the paper industry. However, they had much less knowledge of the specific experience and existing capabilities of Tampella at the time they entered key positions in the company.

The second type of new managers coming to Tampella can be called technical specialists. In contrast to the experienced specialists, most of them did not have extensive previous experience in the paper industry. Instead, their formal training in paper industry engineering was on a par with or above that of their colleagues at Tampella. Because they were young they came to Tampella either straight from the technical university or from their first or second position in another paper industry company. They were therefore originally situated on the lower rungs of Tampella's managerial ladder. In fact, there were several technical specialists, many of whom had only limited experience in the paper industry, but who were now offered challenging jobs in the expanding management structure of Tampella. As a matter of fact, not too many competent engineers were at that time available as most of the Finnish paper industry companies were introducing new technology. Although some of the technical specialists had experience of production equipment similar to that now being acquired by the Tampella paper mill, most of them were technologically inexperienced compared with the typical technology enthusiasts or production functionalists.

The technical specialists were more restricted in their thinking than the experienced specialists. This is to say they did not yet think about strategy but instead considered paper making mainly as a technical process. The technical specialists had learned to know the paper industry as a technologically advanced and progressive branch of industry, which it in fact had been in Finland for several decades. In a sense, the use of advanced technology was what made the paper industry interesting to them. Moreover, because they were young and had moved between paper industry facilities in Finland, they did not have strong bonds to any Tampella localities.

As a consequence of this expansion the constellation of the managerial actors at Tampella was significantly altered. After a short period, there were six distinctive actors instead of four and in addition, one of the previous actors had been replaced. Thus the managerial actors at Tampella had become even more mixed as three qualitatively different new actors had entered through and after the change in corporate ownership. This also means that in the course of the technological discontinuity in focus, the constellation of actors gradually became more diverse and differentiated. The deviating backgrounds and approaches of the six actors to managing paper industry activities are presented in Table 6.2.

We argued above that there were clear conflicts and contradictions within the management of Tampella before the technological discontinuity in focus. In the implementation phase an additional source of conflict was the competition over the managerial positions in the major technological change project. However, as already noted there were also conflicts between the new owners

Table 6.2 Comparison of the background and managerial objectives of the actors at Tampella

Feature	Actor					
	Technology enthusiasts	*Production functionalists*	*Business innovators*	*Risk-seeking owners*	*Experienced generalists*	*Technical specialists*
Age	50–55	45–50	40–45	40–45	40–45	30–35
Education	MSc (Eng.)	Other technical education	MSc (Eng.) MSc (Econ.)	MSc (Econ.) MSc (Law)	MSc (Eng.)	PhD or MSc (Eng.)
Work history	Several positions at Tampella	Several positions at Tampella	One or two positions outside Tampella	Several positions in banking	Several positions at Tampella's competitors	One position outside Tampella
Focus of attention	Production technology	Production process	Products and customers	Corporate value	Corporate competitiveness	New technology
Main objective for the new technology	Modernity	Compatibility with the existing technology	Value added to customers	Increase in the value of the business	State of the art	Modernity

and these old-timers. It is therefore not surprising that there were also conflicts between old-timers and the newcomers. Conflicts emerged partly because the hiring of the latter was initiated more by the new owners than by the old-timers themselves. More importantly, however, conflicts were caused by the fact that the newcomers did not appreciate the old-timers and accept their previous achievements. One of the experienced generalists criticised the Tampella paper machine (PM 3) investment in the early 1980s in the following words: 'The choice of newsprint in that situation indicated a lack of managerial intelligence'. The newcomers also did not like the personalities or management style of some of the old-timers. In the words of Tampella's incoming CEO, some of his new subordinates 'were not real managers'.

Thus the divisions within management did not vanish but instead their sources and forms changed as Tampella moved forward from the initiation of the technological discontinuity to implementation. It should be acknowledged that the deviating objectives and tendencies of the actors are partly related to the differences in both the managerial experience and the actual managerial responsibilities of the actors and in the differences in the personalities of the individual managers and their professional education. Thus the present study does not contend that these differences are in any sense exceptional. However, the fact that in this period the management of Tampella was composed of actors with clearly deviating managerial objectives makes the case interesting from the theoretical angle of this study. Thus it is now important to examine how these deviating actors were mobilised to bring about a discontinuous change in Tampella's paper industry activities. The aim of the following section is therefore to demonstrate how the divergent managerial objectives of the actors were in congruence with the technological discontinuity in focus.

Mobilisation of managerial actors: the discontinuity as a response to divergent managerial and personal objectives

In this section we examine how these divergent actors came to support and work for the same technological discontinuity. In other words, the section explores the mechanisms which brought each actor to support the discontinuity. The first subsection examines the managerial objectives of each actor and the second their more personal aspirations related to the discontinuity.

The managerial objectives of the actors in relation to the technological discontinuity

For the risk-seeking owners the technological discontinuity in focus was justified for several reasons. First of all, it offered an opportunity to use the solution they actually had for solving Tampella's strategic problems: financial resources. In fact, the risk-seeking owners believed that the low performance of Tampella was related to a lack of investment.

From their point of view this problem was now solved because at that time there was no limit to the amount of money available on the international market for profitable use. Making major investments thus raised hopes for improvement in the state of the company. From this perspective the initiated technological discontinuity was a tool to increase the value of the Tampella paper mill. This was also reasonable because though the mill's reputation was not good, its turnover was one of the highest among Tampella's business units. Investing to this facility effectively signalled that although Tampella lacked previous references, it was now aiming for the technological leading edge. Another reason for the risk-seeking owners to support the technological discontinuity in focus was the fact that it was one of the few ideas that the old-timers were able to produce in a reasonable time after the company take-over. This was also one of the reasons for the quick investment decision which also reflected the risk-seeking owners discontent with Tampella's existing management. The technological discontinuity was also credible for non-experts in paper industry like the individual risk-seeking owners because it resembled what the other Finnish paper industry companies had already done or were simultaneously doing. In addition, as Tampella also made machinery for the paper and pulp industry, the acquisition of new machinery at Tampella paper mill might have positive effects for the rest of the company as well.

Although it would not have been easy for the technology enthusiasts to formulate the technological discontinuity in focus, it conformed to many of their managerial objectives. Most importantly, it represented another opportunity to protect the continuity of Tampella's paper production. More concretely, PM 2 which now scheduled for rebuilding was an old machine. It would not last long without major renovation. The concept now developed for the machine was ambitious but the technology enthusiasts nevertheless believed that it had to be done. In the words of an individual manager: 'We just have to learn to coat papers on-line'. In addition to its high potential for profitability, the technology enthusiasts considered the paper machine project at Tampella an exceptional opportunity. They were aware that building new paper machines at Tampella is a rare occurrence and therefore all openings for such projects must be utilised. In fact, the technology enthusiasts had long worked in a company which suffered from a lack of capital and they had also experienced the cancellation of one investment project in the late 1960s. Keeping this in mind, it is understandable that they were also sceptical until the formal investment decision. Consequently, they did not initiate hiring of new managers from outside Tampella.

In contrast to the other actors in the management of Tampella, some of the individual production functionalists did not originally support the technological discontinuity in focus. Those who did appreciated the new technology and believed that all investments in new technology are important prerequisites for further mill development. In contrast, those who originally opposed the discontinuity considered the changes too demanding and technically risky. They were critical towards the potential profitability of the new paper machine (PM 2) because they had experienced the practical difficulties of reaching the theoretical capacity of Tampella's PM 3. They also did not like the idea of breaking up the old PM 2, which also implied losing all that was achieved with the machine through production development work in previous years. In fact, all rebuilds, even the ones in which the new technology has been extensively tested elsewhere, have the potential for disrupting the production process and ruining efficiency. The lack of support for the discontinuity by the production functionalists was also due to the fact that instead of totally replacing the old production technology, new technology could also have been added to supplement the old. From the production functionalists' point of view such changes would have been more compatible with existing technology, production processes and possessed capabilities.

For the business innovators the technological discontinuity at Tampella was a most welcome project. Most importantly, it carried out those managerial objectives that they had imported to Tampella's paper industry management. In brief, the changes implied emphasis on marketing instead of production activities and incorporated a fashionable paper product to replace the previous ones, which were too standard. From the business innovators' point of view, Tampella now had a chance to terminate newsprint production as many of its domestic competitors had already done a few years before. Moreover, the discontinuity meant a shift towards niche-marketing and from producing newsprint for anonymous customers to producing several special paper grades for specific and potentially loyal customers. This was feasible because printing paper customers prefer to choose mill-specific paper grades. This, however, requires the creation of mill-specific contacts with these customers. Consequently, the marketing functions became critical in the management of the Tampella paper mill.

From the experienced generalists' perspective the discontinuity in focus represented an effort to make Tampella's paper industry operations 'state-of-the-art'. They thought that the changes now being implemented at Tampella were wise, despite the fact that some of Tampella's competitors had already carried out similar projects. Nevertheless, the technological discontinuity in focus was something that the Finnish paper industry companies had to do, regardless of their previous experience. The individual experienced generalists were aware of Tampella's difficult past, but they nevertheless believed that a rapid increase in Tampella's performance was feasible if the potential available in the discontinuity were realised. In fact, they were more concerned that the discontinuity might not be ambitious enough. In their view of the world, every new technological change project should exceed the results of its predecessors, irrespective of which company performs it. For the experienced generalists technological discontinuities and the risks entailed were an integral part of paper industry evolution.

The technical specialists considered the discontinuity at Tampella reasonable because it would obviously bring modern technology to a paper industry company that was previously not first-line. Their framework on paper industry operations mainly included categories related to the technical processes and therefore they did not view the discontinuity as a business-strategic act. Instead, the individual technical specialists were self-confident and fascinated by the prospects offered by applications of new technology to the paper industry in general and to the Tampella paper mill in particular. Although Tampella's previous technological standard was not high, the present efforts to make it technologically up-to-date were more important. A short time earlier they would not have been interested in coming to Tampella, as its paper mill had not kept up with industry-level innovations.

To summarise, we examined above how the technological discontinuity at Tampella was related to the divergent managerial objectives of the actors involved. That is to say the study has identified how the actors' objectives conformed with the initiated changes. The main findings in this analysis are shown in Table 6.3. However, this does not exhaust the topic. On the one hand, this is because the analysis is not sufficient to account for how the actors, who had so long accepted unambitious and risk-aversive forms of change, now suddenly supported a highly ambitious technological discontinuity. On the other hand, it is also unclear why the Finnish paper industry 'champions' were suddenly motivated to join the management of Tampella, a long-term technological laggard among its domestic competitors.

In other words, more study is needed to explain the fact that most of the actors involved in the discontinuity took a significantly new position. For example, the technology enthusiasts supported the discontinuity and did not seem to pay much attention to its inherent technological and market risks though over the years they had always preferred low-risk alternatives even in small-scale investments. Neither were they concerned about the sufficiency of earlier capabilities for manufacturing the new high value-added paper grades when supporting the proposed move by Tampella, which had traditionally been a newsprint producer, into the unknown market for magazine papers. It is also peculiar how the production functionalists accepted the discontinuity, as some of them initially responded to it with considerable scepticism. Moreover, it should be noted that the business innovators had the courage to support the discontinuity though they were almost totally unfamiliar with its concrete technological requirements. It is also interesting how the risk-seeking owners dared to approve this major technological change project on the basis of only preliminary plans and how the experienced specialists took the risk of raising the stakes during implementation although they knew Tampella was inexperienced in managing this kind of project.

We believe that the courage necessary to initiate and implement the technological discontinuity must be associated with the personal interests of the actors. Only then is it possible to understand how a long-term technological laggard suddenly aims for the state of the art in the Finnish paper industry. Whereas the analysis above accounts for why the discontinuity was motivated in general, we still have to examine what were the more personal aspirations involved and how they both influenced and were influenced by the implementation of the technological discontinuity. For example, we believe that various personal interests

Table 6.3 Comparison of the actors' relationship to the technological discontinuity at Tampella

	Actor					
	Old-timers			Newcomers		
Trait	Risk-seeking owners	Technology enthusiasts	Production functionalists	Business innovators	Experienced generalists	Technical specialists
Attitude to the discontinuity	Positive	Positive	Complying	Enthusiastic	Positive	Positive
Reason	Adds corporate value	Creates continuity	Threatens the 'learning curve'	Brings products and customers to the focus	Takes Tampella towards the technological leading edge	Uses latest technology
Role in the discontinuity	Provides resources	Provides credibility	Responsible for the technical work	Benchmarks Tampella's competitors	Encourages the ambitiousness of the project	Provides 'new blood'
Problem	Lack of knowledge on paper industry	Lack of state-of-the-art vision	Lack of enthusiasm and specific technical capabilities	Lack of technical capabilities	New context and inability to affect the project design	New context and lack of experience

and aspirations were involved in the adjustments made to the technical characteristics of the adopted technology, the appointments and dismissals of individual managers and the changes in the collective commitment to the project. In other words, understanding the loss of courage and commitment and the later departures of the actors involved requires an examination of their personal goals. The next subsection therefore aims to conceptualise the mobilisation of managerial actors in Tampella by demonstrating why the different actors were first committed to the technological discontinuity and then lost their enthusiasm.

The personal interests involved in the mobilisation

At the time of the formal decision concerning the technological discontinuity, the risk-seeking owners were in an ambiguous situation. On the one hand, the Tampella take-over still offered an opportunity to continue their 'success story' in the industrial ownership business. As a matter of fact, the individual risk-seeking owners had only recently been appointed to their formal positions in the bank largely as a consequence of their success in these operations. This means that the success in the management of Tampella was not a means for further promotions, but was instead needed to legitimate the line of business development which they had promoted at Skopbank. On the other hand, the risk-seeking owners were disappointed with the state of the corporation and the stagnation they perceived. For example, Tampella's balance sheets had promised much more than was justified by concrete evidence. As already mentioned above, they were also disappointed by the small number and poor quality of the investment proposals made by the management of Tampella's business divisions despite deliberate encouragement. The new top managers hired seemed to change this overall picture as some new projects were being prepared. However, the anxieties Tampella's new owners felt at that time are reflected in the following words of an individual manager:

> Right after the take-over, we realised that something had to be done soon. There were two reasons for this; the investments had been few and the depreciation slow. On the other hand, our contacts with the management at the time suggested that they were avoiding action. This was understandable as they were convinced that anything costly was impossible. As a result, they had decided to restrict action in the corporation. But now when the situation was different, it seemed management had only a few ideas of what to do.
>
> (Laurila 1995: 90)

In this situation the new owners were not afraid of taking drastic measures to 'bring Tampella around'. This is to say projects such as the technological discontinuity in focus were not supported only because they had potential or were useful as examples for the whole corporation. Instead, they were also supported because only highly ambitious endeavours were sufficient to ensure an acceptable increase in the value of Skopbank's most important industrial acquisition. This also means that the bankers were somewhat worried about their own personal reputation. The formal responsibility for managing Tampella was a difficult task for the risk-seeking owners. In fact, they were only part-time managers for Tampella as most of their time was taken up by the bank. This combination was workable as long as both Skopbank and Tampella had no major problems. However, as difficulties emerged on both of these fronts during 1989 and especially because Tampella had by this time initiated several new acquisitions and other ambitious change projects, the combination became intolerable. A related problem was that the new hired heads at Tampella also seemed unable to meet the risk-seeking owners' expectations in managing the company.

The technology enthusiasts' attitude towards the technological discontinuity at Tampella largely reflected their previous experience and current position in the managerial hierarchy of the company. Many of the individual technology enthusiasts held established management positions at the time the discontinuity was initiated. The project was therefore not an opportunity for them to advance in their careers. For example, they were not interested in heading the concrete building of the machinery although they had previously headed such projects at the mill. Instead, being involved in the formulation of discontinuity and being formally responsible for Tampella's paper industry activities as a whole they considered the project mainly an opportunity to participate once more in a leading edge project before retirement. In fact, they had become committed to the Tampella paper mill in previous decades, and were proud of their success in rebuilding under the threat of mill closure in the early 1980s. The technology enthusiasts nevertheless felt a moral burden to exploit the opportunities now created, as they had been in charge during some of Tampella's previous failures. For some technology enthusiasts the project also promised to muffle previous criticism of Tampella's paper industry management. As one of the technology enthusiasts put it:

> I've been mixed up with these things for thirty-five years now and for part of that time in such a position that I should have been able to speed up the [investment] decisions [on new technology]. However, I'm glad that I finally had the chance to be involved in making this big change.
>
> (Marttila 1988: 107)

It seems that the technology enthusiasts were not particularly afraid of becoming incompetent after the discontinuity. This is understandable because most of them were not involved with the actual paper production process. However, the initiation phase for the discontinuity was somewhat problematic for them because they were not exactly sure about the state-of-the-art technology and products in the industry. In particular, they lacked knowledge of the characteristics of the paper products and production technologies which would make Tampella's new PM 2 better than the machines of the other producers in this field. However, one of the benefits of the discontinuity for the technology enthusiasts was that it also offered them an opportunity to continue their previous co-operation with Tampella's machinery builders. Such details made it even easier for them to endorse the initiated discontinuity at the critical moments of decision making.

From the production functionalists' point of view the initiated technological discontinuity was in many ways problematic. Because most of them were responsible for the management of the actual paper production process, they obviously risked becoming incompetent for their positions. This was especially so because career advancement at Tampella had traditionally been based more on managerial seniority than on expertise in paper industry management. Although the production functionalists appreciated adding value to Tampella's paper products, the initiated discontinuity was disturbing for them because it posed a threat to the mill's current production efficiency and simultaneously to their personal achievements. Whereas the technology enthusiasts considered the acquisition of the new equipment in the early 1980s their personal victory, the production functionalists felt they had succeeded in exploiting that machinery in previous years. Moreover, in contrast to the technology enthusiasts, the production functionalists were younger and had more to achieve in their careers. Thus the way the discontinuity was formulated and how it would succeed would have significant consequences for them.

The discontinuity was inconvenient, especially for those production func-tionalists who were not appointed to head the concrete project or its various sub-projects. This is not to say that the production functionalists would have typically been eager to reach the upper ranks of the managerial hierarchy. In contrast, most of the individual production functionalists were comfortable on the shop floor. However, by criticising the discontinuity, the individual produc-tion functionalists were protecting their position within Tampella's internal professional market. For example, making minor improvements instead of acquiring major new technology would have kept them from losing their competence.

The fact that the opposing production functionalists nevertheless agreed to the discontinuity after it was formally decided is understandable in terms of their working experience. In fact, both they and the technology enthusiasts had become loyal to their employer during the long period in which they had occu-pied key positions at different levels of the mill management and had advanced in their careers. However, because the experience of the production functional-ists was gained only at the Tampella paper mill and because the mill lacked expenditure in state-of-the-art products and equipment, their self-confidence was low as colleagues employed by competitors at the same time had increased their skills and capabilities. Moreover, because they also lacked the highest training in paper industry engineering, their chances of gaining an equal managerial position outside Tampella were poor. As one manager stated:

It seems that I was not sufficiently aware of what was involved when I took a managerial position and responsibility for the production process. As a consequence, I focused too narrowly on the day-to-day responsibilities of the position and neglected to think about development of the mill in general and my personal professional development in particular.

(Laurila 1997b: 265)

Apart from the fact that the technological discontinuity in focus conformed to the business innovators' managerial objectives, it also carried many personal advantages for them. Most importantly, it offered the business innovators a chance to improve their position in the management of Tampella, which had traditionally been dominated by actors focusing mainly on technical issues. This was because most of the individual business innovators were interested in moving up Tampella's managerial hierarchy although their field of expertise did not provide the best starting point for this goal. Nevertheless, the business innovators had already been able to apply their capabilities in industrial marketing at the Tampella paper mill because there was no one to challenge them. These accomplishments had been a source of pride in a context where technical skills were held in high esteem. In fact, this credibility gap prevented the other actors from considering them capable to assess the technical specificities of the discontinuity, despite the fact that they were the most active group in its formulation.

Apart from the fact that the technological discontinuity provided an opportunity to win recognition for the mill and for Tampella as an up-to-date Finnish paper producer, it also changed the nature of managerial work in the company. From the business innovators' perspective there was much additional work in keeping contact with the customers around the world. Some of the individual managers also liked to travel. The business innovators soon noticed the increase in the amount of work needed in this field and promoted recruitment of managers specialising in sales and product development. In a way, changing the composition of the management at Tampella in this direction further enhanced their position.

The experienced generalists who joined the management of Tampella in the implementation phase of the technological discontinuity had at least two kinds of personal reasons for coming to Tampella. First, most of them obtained more responsibilities than they had had in their previous jobs. The experienced specialists were appointed to key management positions at Tampella, which seemed to be a natural development in their ascending careers. In fact, most of the individual experienced generalists were willing and were accustomed to operating at top management level. They were also even more interested in promoting their careers than for example the business innovators.

Second, the move to Tampella was also an opportunity for the experienced generalists. Although they had been successful in many ways, some of the individual experienced generalists had recently experienced disappointment which increased motivation for their new responsibilities. It can be expected that when a corporation recruits new managers it often has to accept individuals who are not fully satisfied with their current position. For example, the new Tampella CEO had been put under pressure in his previous position, as the business division under his control was taken over by its domestic competitor. Another manager did not receive the appointment he was expecting. The appointments at Tampella provided them with compensation in the sense that instead of being tormented by their superiors they now enjoyed the full confidence and financial backing of Tampella's new owners who did not have the time or competence to restrict their activity. For example, the new CEO explained that the instructions he got from his new superior were as follows: 'Take a good look at the company [Tampella] and give us your view on what needs to be done' (Laurila 1995: 93).

Through their consistent support for the initiated technological discontinuity, the experienced specialists convinced their appointers, who were still uncertain that this was the right thing to do. Because they were unfamiliar with the history of both the mill and the company, it proved easy to adopt the ambitious objectives promoted by the risk-seeking owners. This was also easy because they were able to transfer the high level of ambition from the facilities they had previously managed and thus aimed to repeat their previous successes. This was also one way to control the other actors; they could force the old-timers to react and thus discover who they could most easily co-operate with in the future. Moreover, by fully supporting the ambitious discontinuity at Tampella, the experienced generalists were able to show to their previous superiors that they were not out of the business. The ambitiousness of the experienced generalists is reflected for example in the following public statement by Tampella's new CEO in early 1989:

> An organisation is thriving when it takes on big challenges, preferably too difficult rather than too easy. Then the entire personnel has to develop new solutions, new ways of acting and more efficient ways of planning for the future. . . . We must be extremely upset by every unprofitable business unit. This cannot go on, something must be done and it must be radical.
> (Laurila 1997b: 261–262)

The ambitious visions of the experienced generalists served as an important signal to the technical specialists who were entering the lower levels of Tampella's managerial hierarchy. This was partly because the youngest technical specialists often lacked detailed information on the capabilities of individual mills and paper producing companies. They also knew very little of the specific features of the Tampella paper mill as the context for applications of the new technology. In fact, they were less interested in which mill or company they were employed by as long as the management style and the technology applied were leading edge. Thus it was no surprise that they were interested in being involved with the discontinuity in focus as it was one of the most widely publicised projects in the Finnish paper industry at that time. In addition, Tampella had received much positive publicity after the ownership changes and prospects for careers and opportunities for further development projects seemed to be limitless.

Coming to Tampella was therefore motivated for the technical specialists, especially because they wanted to be involved with new technology which they also considered the basis for boosting their careers. Their self-confidence increased as they were in many cases granted a position of authority whose achievement in the Finnish paper industry usually takes many years. For example, some of the technical specialists were appointed production or project managers almost directly after graduation. In addition to offering opportunities for further learning and career development, the technological discontinuity at Tampella also enabled the technical specialists to show their capabilities to the community of Finnish paper engineers, which they believed to be following events at Tampella with interest. This is not to say that the technical specialists would have been eager to proceed to the upper ranks. In contrast, the fact that Tampella was now dealing with technologically difficult and unexplored areas seemed to be a more important factor for the entering of some individual technical specialists. As one of the individual managers stated:

> The main reason why I came here was the novelty and maybe partly the challenge: how to make an established mill produce a new product range and to solve the many problems this naturally causes. But I felt that the mill needed to be able to face the demands of manufacturing coated paper grades.
>
> (Laurila 1997b: 266)

To summarise, in the above we examined the personal interests of the actors and how they were realised in the technological discontinuity at Tampella. In other words, we described both the actor's managerial objectives and their more personal interests and aspirations in relation to the technological discontinuity in focus. Most importantly, the analysis shows how such divergent actors found themselves supporting and working for the same major technological change project.

The analysis demonstrated that actors from different backgrounds may use technological discontinuities as tools for promoting their various objectives. This is to say that the almost unanimous support for the technological discontinuity at Tampella was based on the good match between the characteristics of the discontinuity and the objectives and interests of the actors involved. This is not to say that the discontinuity would necessarily have the same meaning for each actor. In other words, the actors supported their personal conception of the discontinuity. Moreover, some managers supported the discontinuity more on the basis of their formal responsibilities, whereas other managers also had substantial personal commitments.

In this case, mobilisation is indicated both by the short time in which both the initiation and implementation of the ambitious technological discontinuity took place and by the expressed commitment and enthusiasm of the individual managers involved. In fact, all actors, excluding the production functionalists, were inspired at least for a short time by the discontinuity and the vision it opened for their personal future. For example, the new Tampella CEO expressed his feelings at the early stages of the implementation of the technological discontinuity as follows: 'Now I can say that I really enjoy my work. I have freedom to activate my ideas and I have enthusiastic people supporting me' (*Riskienhallinta* September 1988: 16). The same excitement is also reflected in the following statements by an old-timer and a newcomer in the management of Tampella:

> It seems that expectations of the actual changes in relation to both timetable and operational performance were overly optimistic. But at the time it all felt realistic and we went a bit crazy. The situation was so unusual that although we should have been able to restrain ourselves, we just all believed in it. . . . In the initial stage of project implementation the spirit of action within management was high. However, the expectations of what the change would bring and what each one should have been doing were not quite clear.
>
> (Laurila 1997b: 264)

However, as can be expected on the basis of the literature examined earlier in this book, such mobilisation of managerial actors is only a temporary phenomenon. This is comprehensible because the involvement in managing technological discontinuities is until the concrete implementation phase mainly based on expectations. When the outcomes of the discontinuity are gradually exposed the momentum created in the mobilisation inevitably declines. The previous chapter of this book described that the technological discontinuity at Tampella failed to meet many of its official short-term objectives. The background of the mobilisation decline in this case is, however, much more complicated. This is especially because the relationships between the different actors in the actual process of change greatly influence the endurance of mobilisation. In the case of Tampella, the intra-managerial conflicts that emerged in the implementation phase of the technological discontinuity between 1989 and 1990 virtually ruined the previously established coalition of actors. These issues are examined in the following section, which demonstrates the basis for the decline of mobilisation by describing the impediments for each actor's long-term commitment to the initiated discontinuity.

The decline of mobilisation: the discontinuity as a threat to the actors' objectives

The decline of mobilisation in the management of Tampella can be described as a sequential process in which the failures in the implementation of the discontinuity and the conflicts between actors produced a gradual loss of commitment to the project. In other words, the loose coalition of old-timers and newcomers did not hold together. The details of this process from the perspectives of each actor will be examined below.

The risk-seeking owners' loss of commitment to Tampella in general and the major technological change project at Tampella paper mill in particular was mainly related to their failure to meet the owners' expectations. After feeling for a short while that the project might succeed, the risk-seeking owners soon found that their investments were not yielding the desired return. This was not a total surprise in the sense that although they encouraged all actors involved in the technological discontinuity, the new owners had nevertheless come to believe that the project could not be successfully managed with the existing managerial staff, who they felt continued to work as they had under the previous owners. The replacement of Tampella's CEO signalled this and as a consequence, the coalition of managerial actors formed during preparation for the discontinuity began to break up before the project got under way.

However, to the risk-seeking owners' surprise, neither were the new managers, who the owners believed to be competent, able to solve Tampella's problems. After realising that the technological discontinuity in focus and also some other projects had run into problems, the owners no longer allocated resources to new projects. Instead, they started to search companies willing to buy Tampella as a whole or in small parts. They therefore radically altered their expectations for Tampella and the actors involved by withdrawing their support from the company in general and the project at the paper mill in particular. The owners' disappointment with the events at the Tampella paper mill is reflected in the following words of an individual manager: 'We were under the impression that it [the new PM 2] would run at full speed after some initial problems. The start-up, however, was a big disappointment' (*Talouselämä* 1 February 1997: 24).

For the technology enthusiasts the process of losing commitment was somewhat different. After first being pleased to find out that the paper product business they had so long promoted at Tampella was now being funded, they soon discovered that they were not trusted by the new owners or by the managers appointed by the new owners as their superiors. Some of the technology enthusiasts became aware of this when they were appointed to staff positions and new managers with different backgrounds and rhetoric took over their previous positions. In a sense, the incentives which motivated the incoming managers demotivated the managers originally responsible for the technological discontinuity. One of the technology enthusiasts expressed it as follows: 'The newcomers' style in project management deviated completely from Tampella's previous tradition; they criticised the adopted project design and insisted that the mill management play a more active and ambitious role in making the changes' (Laurila 1995: 97). As a consequence of being replaced, some technology enthusiasts only could follow the happenings from the side. Those technology enthusiasts who remained in key positions in the project disliked the gradual changes made by the newcomers in the original technology concept. Moreover, the consultants with whom they had worked in their previous jobs were replaced by new ones, who were considered more suitable by their new supervisors. Thus the project could not be managed in the way the technology enthusiasts would have preferred.

Although some of the production functionalists originally opposed the technological discontinuity, most of them had important responsibilities in its implementation. Some of them headed the project and some the operations of the rest of the Tampella paper mill. They therefore had to face most of the technical difficulties inherent in building the new machinery. The contradictions between the technical demands and the high performance expectations resulted in conflicts between individual managers who had previously been able to co-operate with each other. Another problem for the production functionalists were their new supervisors, who wanted to redefine the technical characteristics of the new paper machine; contracts with technology suppliers had therefore to be renegotiated. The individual production functionalists' commitment was also diminished by the fact that although they did their best to manage the difficult project, their new supervisors blamed them in public for declining performance at the mill. Replacements of key individual managers made when the difficulties at the mill were most apparent were the final blow. In the words of an individual manager: 'Replacements of individual managers placed the blame for the failures on them' (Laurila 1995: 99). After being transferred to less critical positions some of the individual production functionalists left Tampella.

The business innovators did not lose their commitment to the technological discontinuity in focus as soon as the other old-timers. This was because only one of them was replaced in early stages of the project and some were actually able to keep their positions throughout the project. Moreover, the modifications in the original technology concept were related to the characteristics of the machinery and not to the products with which the business innovators were more involved. Neither were they concerned about the technical difficulties involved in building the new machinery. Instead, some of them had enjoyed the opportunity to travel around the world telling about the forthcoming product.

However, at its late stages the project also started to threaten the business innovators' prestige. They found it unpleasant that the new product they had been advertising to individual customers was to be significantly delayed. Their motivation was diminished even more by the fact that key business innovators who had initially proposed the discontinuity and spoken in favour of it were removed from the mill's management. The discontent of both the business innovators and the other old-timers was increased by the fact that many of the newcomers entered Tampella in groups which had already been working together outside the company. To them it seemed that the newcomers co-operated more readily with each other than with their new subordinates. Moreover, when new managers with little experience in the field were abruptly given major responsibilities, some considered this the beginning of a new form of favouritism. From an individual business innovator's point of view,

It seemed that the newcomers were able to accept only their previous friends and colleagues as their subordinates. If all of them had been experienced and competent they probably would have gained the confidence of the others. Now there emerged a kind of first and second floor effect within the management.

(Laurila 1995: 98)

The frustration of the previous heads at Tampella is also indicated in the following comment on the new hirings: 'A large number of these new managers entered both company headquarters and lower management levels. "Brains" were hired for Tampella and for example division staff got some new people, but those positions were too demanding for them' (Laurila 1997b: 263). In fact, displacement of the business innovators led in many cases to their departure from Tampella. Thus the individual managers who had initiated the discontinuity were unable to manage it to the end.

At the outset, the experienced generalists were not as highly committed to the technological discontinuity at Tampella as for example the business innovators. This was because most of them were hired when the project had already been running for some time. However, the commitment of the experienced generalists increased after they made some changes in the characteristics of the new technology. For example, instead of being similar to some of the machines of Tampella's competitors, the new paper machine (PM 2) was now to be the most efficient paper machine in this product category. The experienced generalists could also increase the pressure on other managers by promoting the idea that if the rebuild was to be successful, it must be only one link in a long chain of technological discontinuities, new machines and modernisations gradually leading to technological superiority. Another factor increasing the commitment of the experienced generalists was the opportunity to hire some of their previous colleagues. However, the appearance of the problems at the Tampella paper mill led them to stress that they had not been employed by Tampella at the time the discontinuity was initiated.

One effect of the emerging problems at Tampella was that they legitimated replacements of Tampella managers both those already made and those forthcoming. In fact, some of these replacements were delayed because the project managers had become practically irreplaceable. They were the only persons familiar with all the innumerable details of the discontinuity. One of the experienced specialists explained the situation.

The managers responsible for the rebuild project were not getting on well with each other. They exceeded their capabilities in many senses and were not able to trust each other or to accept help. Thus when difficulties emerged the blame was mainly put on the others. These problems were impossible to resolve because 'the train had already left the station' and only these few persons knew the situation and were able to protect the mill's interests in relation to the technology suppliers.

(Laurila 1995: 94)

The experienced generalists nevertheless initiated replacements especially because they felt that many capabilities were still missing. One manager gave the following reasons for the replacements:

> Many of the old-timers were plagued by a kind of weakness and lack of self-confidence in relation to their superiors. This was indicated by their enthusiasm for the new opportunities, which nevertheless were not followed by concrete actions. On the other hand, in many cases it seemed that the managers just lacked the necessary skills.
>
> (Laurila 1995: 96)

They also had the support of the risk-seeking owners for the replacements. This does not mean that the replacements were easy to make. One of the experienced specialists described the situation:

> The challenge the company [Tampella] was facing required a new managerial composition. The old guard was unable to cope in that situation and to make their subordinates accept the new style of action. At different levels of the managerial hierarchy there were managers who reacted critically to all change and prevented the emergence of a positive attitude towards ambitious objectives. The problem was that after displacement of a manager who had been liked, there was usually more opposition at the level below him.
>
> (Laurila 1995: 97)

However, because the old-timers resented many of the changes the experienced generalists initiated, the latter became more and more suspicious of the old-timers' will to co-operate and reported their criticism to the owners in order to legitimate hirings of new managers from outside.

The fact that the new owners ended their unconditional support for Tampella in 1990 was a major threat, especially to the experienced generalists. They had come to Tampella not only for what was now going on but also for future endeavours. Moreover, being formally responsible for a poorly performing company or division put some individual experienced generalists into a difficult position. This was not the least because the experienced generalists had originally enjoyed considerable confidence from the risk-seeking owners. The owners had personally hired these managers who then were openly optimistic about the time needed to turn Tampella around. For example, the experienced generalists originally accepted the content and timetable for the technological discontinuity in focus.

The technical specialists were originally enthusiastic about the technological discontinuity at Tampella. The enthusiasm and self-confidence of the technical specialists encouraged the experienced generalists to provide them with new responsibilities. In fact, some of the technical specialists were eager to show their capabilities after their previous supervisors, whom they were unable to appreciate, had been removed. They also supported the alterations in the original technology concept and believed that various technical obstacles would soon be conquered. However, the youngest technical specialists were apparently not at all realistic in their aims. For example, they felt they could soon surpass the previous production records at the Tampella paper mill, although they lacked the support of shop-floor personnel. They also had the courage to modify the composition of paper coating ingredients on the new paper machine (PM 2) before it had reached full production. As a consequence, there were failures that exhausted individual technical specialists and led to their eventual replacement, sometimes by members of the old staff. Some technical specialists also left voluntarily soon after it was clear that Tampella would not launch major technological change projects in the near future.

The fact that there had been public boasting about the success of the discontinuity beforehand made the shortcomings seem even worse to the technical specialists. One of the managers involved described the situation in this way:

> At first, the negative publicity and the replacements of project managers were depressing. But the personnel soon became numb to it and unresponsive to all communication from outside. In fact, everything happened rather quickly. First came the positive publicity which put the mill personnel under extreme pressure as they faced the first problems. Soon thereafter came the negative publicity.
>
> (Laurila 1995: 99)

In summary, during the implementation of the technological discontinuity at Tampella the original enthusiasm and commitment of the involved actors faded. The failures at Tampella justified replacements of several managers, thereby eliminating a significant part of the knowledge gathered during previous decades. As no actor felt completely responsible for the discontinuity, the pressures were revealed in mutual accusations and complaints. One manager described the situation as follows:

> More problematic than the actual technical difficulties was the quarrelling they caused within management. It became more important to argue about who was responsible for each failure than how they would be overcome. In that situation the actors loudly questioned each other's competence.
>
> (Laurila 1995: 98)

Later the conflicts became even more open because all the actors still had much at stake. The risk-seeking owners were concerned about their reputation and the success of the turnaround they had initiated. The actors at the lower levels in the management chain became worried about their reputation and professional authority.

After examining the actors, the actors' managerial objectives and more personal interests and aspirations, both in the upswing and downswing phases of the mobilisation of the technological discontinuity at Tampella (see Figure 6.1), the study will now elaborate on the analysis. The final section of this chapter aims to sum up the principal features of this specific mobilisation and the factors which seemed to be its key triggers and obstacles.

Triggers for and obstacles to mobilisation for technological discontinuities

The analysis presented demonstrates the nature of management mobilisation as social forms in which the actors' managerial objectives and personal goals coincide to create temporary collective commitment in support of an ambitious technological discontinuity. Thus discontinuous change is motivated by the potential benefits offered to the actors involved. In a sense, conflicting actors may be mobilised as the discontinuity offers many opportunities, although not all of them are likely to be realised. In this section we briefly highlight some of the mechanisms which seem essential both for the mobilisation of managerial actors in this case and more generally.

Mobilisation as a situationally conditioned phenomenon

The case of Tampella shows the importance of the advent of a new dominant managerial actor for the mobilisation process. The change had a dramatic effect on the other managerial actors at Tampella because many things that had previously been considered positive were now considered negative. For example, anything costly was seldom permitted under the previous ownership; now costs were not a problem if the prospects were good. In fact, it can be argued that the commitment and encouragement of the risk-seeking owners broke the barriers which could have blocked the emergence of support for the ambitious technological discontinuity. In particular, their ability to make quick decisions differed from Tampella's earlier management tradition and compelled the other actors to

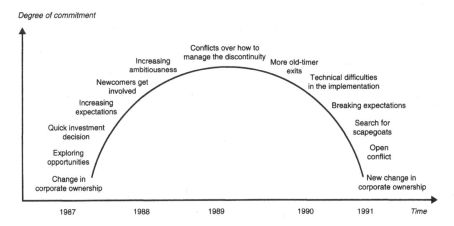

Figure 6.1 Life-cycle of the mobilisation of managerial actors at Tampella

adopt a new mode of action. The change in ownership is reflected in the argumentation of three successive Tampella CEOs in 1980: 'It's up to us, the company management, whether we can employ people in a way that secures profitable production. And of course, all lines of business will eventually show a profit' (Laurila 1997b: 261), in 1986: 'It seems that Tampella has left the worst times behind and we are now making progress. Hopefully, it will already be apparent in this year's performance' (*Metsäteollisuusuutiset* 1986 March: 2) and in 1989: 'Every division of Tampella will end up being the market leader in their business area. To reach that objective we must create growth together and no other thoughts can be allowed' (Laurila 1997b: 262).

The action of the new owners was regarded as both compelling and highly rewarding. This was because the other actors' experience of their sense of worth was now rapidly transformed. For example, the paper mill – which had long been viewed as a burden to the company – had abruptly become a strategic opportunity. More concretely, the board – which had previously concentrated on rejecting all proposed capital expenditures – was now interested in the mill and encouraged all efforts to improve its performance. The message received by the actors at Tampella was that the company had potential. This open encouragement also functioned as an important signal for actors to enter Tampella from outside; without expectations of forthcoming investments 'the paper industry champions' would not have been attracted to the company.

However, the mobilisation of managerial actors again proved fragile. In fact, both the new owners and the other actors started to break with the coalition after the first concrete difficulties and failures were experienced. Thus the actors reacted to a concrete situation in which the potential threats were now emphasised instead of the potential benefits. This is not to say that the situational context completely determined managerial action, because the motivation of an individual manager is also based on his personal experiences, professional background and career aspirations. For example, an individual manager might lack the courage or a concrete skill to take advantage of a situation which for another manager is a sufficient signal for action. However, it can be argued that the mobilisation of managerial actors is always situation-specific in the sense that the situations influence who become mobilised on the basis of both their managerial objectives and their personal interests and aspirations. This means that the temporary resource abundance brought about by the new owners of Tampella in the late 1980s is an example of a situation which permitted both these aspects to coincide.

The conflicting experiences of locals and cosmopolitans

Whereas the change of the dominant actor functioned as a trigger for mobilisation at Tampella, the differences in the experience base between the old-timers and the newcomers functioned as a major obstacle to the continuity of the mobilised coalition. In other words, the conflict was between locals and cosmopolitans (Gouldner 1957; Merton 1957; Gouldner 1958). In brief, whereas locals are loyal to one company and try to serve its interests, cosmopolitans eagerly change from one company to another especially if this permits them to promote the interests of their profession. In brief, competence for cosmopolitans is more important than hierarchy, while the reverse is true for locals. The interface between cosmopolitans and locals is therefore always difficult because the bases of their commitments are different. In the case of Tampella, the self-confidence of the newcomers was based on their perception of superiority in relation to the old-timers, whereas from the old-timers' point of view the newcomers did not know how to operate in the local context. For example, the cosmopolitans did not know the specific strengths and weaknesses of the personnel. This is indicated by the following unpublished statement by a shop steward: 'When an employee gets to work early in the morning, he leaves clear tracks in the snow with his feet. When a company manager goes to work, he leaves no tracks'.

In contrast to the newcomers, the old-timers did not have to earn the confidence of the locality and to become familiar with its specific experiences. The difficulties in crossing this frontier make some conflicts in the implementation of the technological discontinuity understandable. It seems that experience was not fluently transferred effectively between actors who needed to co-operate. One manager described it:

> Knowledge is not necessarily transferred from one generation to another. Often new managers do not know what has been done before or how the existing systems operate and start to build new ones and test many ideas that have proved to be dead ends many times before. And the old staff may not want to tell the newcomers anything unless they are specifically asked.
> (Laurila 1995: 103)

The Tampella case shows both the power and the limits of ambitious technological discontinuities as catalysts for mobilisation of managerial actors in the Finnish paper industry. On the one hand, it seems that acquisition of new technology is a sufficient trigger for migration of top managers from the leading firms to the laggards. This is related to the dominant position of engineers in the management of an industry characterised by challenging technological rebuilds. On the other hand, ambitious technological discontinuities do not necessarily permit mobilisation even within the context of the Finnish paper industry. This is because the views concerning the acceptable level of ambition may vary significantly even among managers representing the same profession. For example, one of the subordinates of Tampella's new CEO commented on his superior who had training and experience in the same field: 'I knew the new CEO well beforehand and was already certain at the time he entered that our views on managing Tampella were contradictory'.

In the case of Tampella paper mill, discontinuities formulated on significantly different levels of ambition would still have been considered ambitious in terms of the mill's previous history. On the other hand, technological discontinuities are always ambiguous in the sense that they permit several different modes of action in accordance with more general managerial objectives. This is to say each formulation of the technological discontinuity allows several alternative approaches to implementation.

The evidence presented demonstrates that competing alternatives for developing production technology imply differing levels of risk and that in fact, there are no objective criteria for choosing many of the technical details related to paper machine rebuilds. Moreover, even within the engineering profession alone there may be different schools of thought concerning for example on-line or off-line coating processes. This means that ambitious discontinuities permit mobilisation of managerial actors, but commitment to these endeavours is restricted by the variety of means used in achieving them.

Management as a constellation of divergent actors and micro-mobilisations

The concept of mobilisation in this study refers to management as a combination of conflicting actors and several simultaneous micro-mobilisations and change programmes. It was demonstrated above how the technological discontinuity at Tampella was produced by a coalition of divergent actors. Thus the managerial coalition that emerged was not unified, but was instead a coalition of actors who supported the discontinuity for different reasons. For example, the technology enthusiasts emphasised the modernity of the acquired technology and the business innovators the need to upgrade products. Now a major improvement in both of these issues was possible at the same time.

For the further development of the concepts developed here it is essential to note that the actors not only supported the discontinuity, but also found each other in several smaller coalitions. To demonstrate this it is useful to make a distinction between the means and objectives of the actors involved in the discontinuity. The mobilisation at Tampella as a constellation of four micro-mobilisations is illustrated in Figure 6.2. We therefore argue that the actors taking part may be united on some issues and divided on others. For example, the technical specialists were not greatly interested in the long-term effects of the discontinuity. For them securing up-to-date technology for Tampella was the key factor. Thus the technical specialists supported ambitious means though in many senses their managerial objectives were not ambitious. However, they readily co-operated with business innovators and experienced generalists who also considered it important to apply sophisticated technology at Tampella. The mobilisation of managerial actors can therefore be seen as a constellation of four micro-mobilisations; three of them promoted the discontinuity and one opposed it.

After examination of the internal structure of the mobilised coalition of actors, the conceptualisation can be further developed by examining the external connections of the actors involved. Then it may be noted that the different actors described above were connected with the specific change

Means

Non-ambitious	**Survival game** Representatives: Irresolute owners Production functionalists Content: Trying to scrape out a living with few resources and risk-aversion Both means and objectives important Emphasis on the short-term	**Mean-different ambition** Technology enthusiasts Risk-seeking owners Aiming at an increase in long-term performance one way or another Dominance of ends over means Emphasis on the long-term
Ambitious	**Short-sighted new technology application** Technical specialists Aiming at using the latest technology regardless of the general objectives Dominance of means over objectives Emphasis on the short-term	**Ambitious turnaround** Business innovators Experienced generalists Aiming at ambitious objectives using the latest technology for the discontinuity Both means and objectives important Emphasis on the long-term

Non-ambitious   Ambitious

Objectives

Figure 6.2 The mobilisation at Tampella as a constellation of four micro-mobilisations

programmes which they were promoting inside or outside Tampella. For example, the technology enthusiasts had been developing liaisons between the forest industry and Tampella's Pulp and Paper Machinery Division, which had been the principal force directing capital expenditures in the company during previous decades (see Chapter 7 of this volume). The production functionalists were part of the community of Finnish paper engineers, a community which had sought to promote production processes through an exchange of technological information and practices across company boundaries and therefore to make minor technical improvements. On the other hand, the business innovators had long built up ties with the emerging community of general managers, a community which had sought to promote the position of marketing and customer relations in the Finnish paper industry, which was traditionally dominated by production engineers. It can also be argued that the risk-seeking owners were participants in a larger mobilisation occurring on the international finance markets which allowed structural reorganisation of Finnish industries during the 1980s. The experienced generalists represented a mobilisation aimed at transforming the entire Finnish paper industry from a process-oriented bulk producer to a special product and customer-oriented high-tech industry. In sum, these simultaneous micro-mobilisations and change programmes were among the resources included in the mobilisation at Tampella.

The perspective above emphasises the diversity of the resources and actors involved in and connected with the mobilised coalition. The study should also determine the limits of mobilisation, that is to note who did not become involved. As a concrete example, the actors at Tampella had previously co-operated closely with the supplier of the company's widest paper machine (PM 3), that is with the Valmet Corporation. Now the fact that Tampella was supplying the new paper machine (PM 2) prevented this co-operation from continuing. Another reason for this was the fact that the new Tampella CEO, who had previously headed one of Valmet business divisions, forbade any co-operation with his previous employer. This meant that the actors who were mobilised partly determined who else could be mobilised. It also seems that the persons in the key roles of the discontinuity may have kept some other managers from taking an active part.

The analysis of the mobilisation for the technological discontinuity at Tampella has so far operated mainly at the level of the managerial actors involved. This does not imply that technological discontinuities should be seen as mere reflections of the actors' intentions. In contrast, the study argues that technological discontinuities represent intentional action which, however, is restricted by its context. This is to say managerial actors use the resources around them to produce technological discontinuities. After having analysed how the actors operate with resources, the book will then elaborate how these resources are created. To demonstrate this in the specific case of Tampella and the Finnish paper industry more generally, the next chapter explores the background of the technological discontinuity analysed in the previous chapters.

7 Building a contextual explanatory framework for technological discontinuities

The previous chapters first identified a technological discontinuity in an established firm and then analysed how this discontinuity is first initiated and then implemented by the managerial actors involved. In other words, besides analysing the details of the discontinuity the study has conceptualised the mobilisation which permitted it to take place. This means that the study has already provided a voluntaristic account for the technological discontinuity in focus. In contrast, this chapter examines the background of the discontinuity and the related management mobilisation in order to provide a contextual explanation for the phenomenon. In this chapter we therefore do not search for generic explanations, but aim instead to capture the conditions which allowed managerial actors to produce technological discontinuity in this specific case. The contextual explanation developed here assumes that technological discontinuities are caused by interaction between the intentions of the actors and the contextual features of the situation. This is illustrated in Figure 7.1. The contextual features direct the actors' attention and either encourage or discourage them to initiate and implement major technological change projects. In this spirit, the chapter will show that the emergence of the technological discontinuity at Tampella is the result of several situational factors.

To reach these aims, the first section of this chapter first identifies the main technological capabilities in the paper industry and then goes on to examine how they can be developed. It shows that technological discontinuities are an elementary part of capability development in the paper industry. After that it

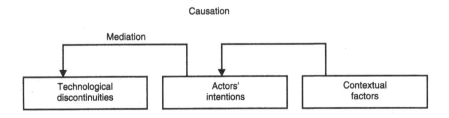

Figure 7.1 Intentional and contextual factors conditioning and contributing to technological discontinuities

proceeds to elaborate these forms of capability development by comparing several major technological change projects performed by Tampella during its history. The analysis will show that although the technological discontinuity examined above was in many ways genuine for Tampella in some other ways it also continued Tampella's traditional approach to the development of its technological capabilities. This means that we should no longer be surprised that such a project was undertaken. Instead, we become interested in the origins of the material resources needed to accomplish such a project and why they were directed to this specific endeavour. The second section of this chapter therefore describes how Tampella's finances changed abruptly in the late 1980s and the third subsection analyses the characteristics of Tampella's business environment, which motivated use of the funds for this purpose. The final section of this chapter draws conclusions on the constructed explanatory framework for technological discontinuities and management mobilisation.

Developing capabilities through technological discontinuities

Development of technological capabilities in the paper industry

Technological capabilities can be defined as a set of differentiated skills and complementary assets that provide the basis for a firm's continuing operation in a particular business (cf. Leonard-Barton 1992: 112; Teece *et al.* 1994: 18). Using this definition, it is argued here that the companies producing paper need two general types of technological capabilities. The first one of these is the ability to build the technology used in paper production. Even if the technology is acquired from outside, the paper producing company must be able to specify the type of technology adopted and to manage the concrete building project. This means that the development of technological capabilities necessarily involves management of technological discontinuities.

In case the technology is delivered – even partly – by the paper producing company itself, the companies need both sufficient engineering facilities to build such machinery and the capabilities acquired through previous machine deliveries. Major technological change projects can be used to develop technology building capabilities at two separate levels (see Argyres 1996a; 1996b). The broadening of technology building capabilities implies moving gradually from the technology needed in a specific part of the paper production process onward and transferring the resulting project capabilities to the forthcoming projects. For example, later in this section it will be shown how Tampella gradually increased the number of forest industry products (e.g. paper grades) which could be produced by its machines. In contrast, deepening of capabilities implies developing more efficient technology for the same part of the paper production process. For example, throughout its history Tampella developed more efficient ways to manufacture the same forest industry products.

The second general type of technological capabilities in paper industry is related to the long-term operation of the production facilities. That is, the paper producing companies need to make efficient use of the machinery during its life-cycle. Developing this capability usually begins when the supplier of the technology starts up the new machinery. The production output is thereafter gradually increased by the paper makers themselves. One way to utilise and enhance these technology operating capabilities is to transfer obtained improvements in one part of the production process to its other parts. The relationship between such operational capabilities and major technological change projects is less evident than in the case of technology building capabilities. However, such endeavours are also essential here because the two main technological capabilities mentioned above are entwined in the sense that even minor developments in paper products usually require building of new production technology. Major changes in paper products are usually prevented by standardisation. Most innovations in the production technologies have therefore been directed to improving the efficiency of the production process (i.e. the widths and speeds of the paper machines).

The relevance of these specificities of paper production technology is that they have encouraged the paper producing companies to adapt their technological infrastructure. In a global perspective this means that these companies have extended their activities to a wide array of businesses between forestry and high value-added paper products such as converted papers (see e.g. Globerman and Schwindt 1986). A typical characteristic of the Finnish paper industry companies is that they have expanded their activities more in the middle of this continuum than at its ends. The Finnish companies have become the largest paper and pulp producers in Europe, mainly by expanding the capacities of their production units through major technological change projects. Another, more relevant consideration is that until recent times many of the Finnish companies have been involved both in the manufacturing of the paper industry machinery and the paper products. On the global level, these capabilities have normally been conducted in separate firms, but in the case of the Finnish paper industry these functions were largely performed by the same companies until the late 1980s. That is, all the main Finnish paper producing companies had their own engineering facilities which took care of mill maintenance and built a significant part of the new machinery.[1] However, whereas the Finnish paper industry corporations have mainly treated their engineering facilities as complementary assets for the forest industry businesses, Tampella systematically sought competitiveness from the interface between these two businesses for over a century.

It will be shown below how successive major technological change projects gradually made Tampella a producer of a variety of forest industry products. Between the middle of the nineteenth century and the early 1990s, as a consequence of the initiated technological discontinuities, Tampella became a prominent manufacturer of paper and board machines, pulping machines and machines needed in supplementary chemical processes and energy generation. Tampella also became a producer of mechanical pulp, board, cartonboard, folded boards, board packages and containers and uncoated and coated printing papers. The following analysis of this development focuses only on the most central events in which successive technological discontinuities resulted in new forest industry businesses. In other words, the events selected reveal a distinctive pattern of diversification through major technological change projects.

Moving Tampella from machine making to forest industry businesses

The relevance of Tampella's long-term technological development

There are at least three reasons for analysing Tampella's long-term patterns of technological development. First, such an analysis supplements material in an earlier part of this book on the relationship between the technological discontinuity in focus and Tampella's previous competences and capabilities. By demonstrating what kinds of technological change projects the company had previously completed, the analysis further develops our understanding of the most distinctive characteristics of the project studied in depth. Second, such an analysis also demonstrates that the technological discontinuity in the late 1980s was in many ways similar to most of the major technological change projects Tampella had performed. In fact, the technological discontinuity in focus is one of the many in which the company's capabilities in the manufacture of forest industry machinery were used to expand into forest industry businesses. In order to demonstrate this fact, we show how Tampella came to use its original capabilities in machine building to supply its own mills with equipment.

In addition to clarifying both the distinctive and ordinary features of a specific technological discontinuity there is also a third reason for taking a long-term perspective on Tampella's technological development. This is because such a viewpoint captures a distinctive pattern of technological change which is typical for the Finnish paper industry sector but which is neglected in the previous management literature in general and in the literature on the management of forest industries (e.g. Globerman and Schwindt 1986; Davis *et al.* 1992; D'Aveni and Ilinitch 1992) in particular. As a consequence, previous management literature lacks conceptual advancement which could be obtained by the examination of a specific type of technological development, namely one in which the producer of forest industry machinery expands its activities into a wide array of forest industry businesses.

As noted earlier in this book, one of the distinguishing characteristics of the Finnish paper industry sector is that it has worked successfully in both the paper machinery and paper products businesses.[2] In this context, Tampella is a unique case because it has, for a period of over 120 years, systematically expanded these two significantly different although technologically related businesses within the same company. Moreover, instead of acquisitions and mergers, Tampella moved on the whole organically from forest industry machines to the production of forest industry products. That is, it built and rebuilt the new production facilities on its own. Thus Tampella offers an opportunity to examine how the technological links between two separate businesses are realised in the long-term evolution of a forest industry corporation.

Seven critical episodes in Tampella's technological development

A number of successive major technological change projects gradually converted Tampella into a highly diversified paper industry company. The main events of this development are presented in Table 7.1. In brief, the analysis here focuses on seven critical episodes in which the company used its paper machine building capabilities to diversify into paper product businesses. The first of these episodes started in 1861, when Tampella was founded as the result of a merger between an engineering works and a linen mill. The former had supplied the production technology to the latter and the facilities themselves were located side by side.[3] Most essentially, however, Tampella had the capacity to make different kinds of machines. In the late 1860s Tampella started to use this capacity to produce machinery for the emerging forest industry. More concretely, Tampella started to deliver wood-grinding machinery to the mechanical pulpwood mills being built in different parts of Finland. An interesting element in this development is that Tampella did not, however, only broaden its capabilities to include the production of paper industry machines, but also simultaneously extended its operations into the forest industry businesses.

As a first step in this long series of technological discontinuities, Tampella supplied the machinery for the new mechanical pulpwood mill which it founded in 1872. As a consequence, in addition to being an important producer of wood-grinding machinery, Tampella also entered the end product business for that machinery (i.e. the production of mechanical pulp). This process continued a little later, in 1887, when Tampella acquired a mechanical pulpwood mill and a board mill. Because Tampella now manufactured pasteboard out of mechanical pulp, it had actually moved one step further in its diversification from forest industry machines to forest industry products. This process proceeded again a decade later in 1897 when Tampella started to produce cartonboard. Tampella had thus moved in three successive steps from the manufacture of pulping machinery to the production of more and more refined forest industry products. This development was based mainly on broadening and deepening its technological capabilities and on the many major technological change projects at Tampella's own forest industry facilities. For example, thanks to the invention of thermo-mechanical pulping, Tampella started the production of thermo-mechanical grinding machines in 1906 and the first two machines of this kind were supplied to Tampella's own mills a year later.

The second of the seven critical episodes comprised a combination of Tampella's different technological capabilities in the supplementary forest industry businesses. More concretely, Tampella's engineering works had already begun the manufacturing of turbines in 1871. However, it was not until 1922 that Tampella used this capability to extend its operations to power generation.

Table 7.1 The process of Tampella's expansion from machine production to forest industry products

Product (year)	End product	Refined product	More refined product
1 Wood-grinding machine (1869) →	Mechanical pulp (1873) →	Pasteboard (1887) Cartonboard (1897)	Cartonboard products (1940)
2 Thermo-mechanical grinding machine (1906) →	Thermo-mechanical pulp (1907) →	Newsprint (1938)	
Power turbines (1871) →	Electricity (1922)		
3 Mechanical pulp mill based on recycled fibre (1945) ⟶		Recycled cartonboard (1945)	Recycled cartonboard products (1955)
4 Semi-chemical pulping (1952) →	Fluting-carton (1962) →	Corrugated cardboard (1962) →	Corrugated cardboard packages (1962) and containers (1964)
Cartonboard machine (1953)			
5 Recycled pulp →	Folded cartonboard (1965) ⟶		Refined folded cartonboards (1965)
Cartonboard machine (1965)			
6 Pressure grinding machine (PGW) (1978) ⟶		Folded cartonboard and printing papers (1983) →	Refined folded cartonboards (1983)
7 Paper machine (1987) →	Coated printing papers (1989)		

Source: Lodenius 1908; von Bonsdorff 1956; Teerisuo 1972; Seppälä 1981; Björklund 1982; Tampella Annual Reports.

This took place when Tampella delivered the turbines for its hydro-electric power plant. A little more than a decade later Tampella began to use the increased amounts of energy generated to produce paper out of mechanical pulp. This means that it took some seventy years to move Tampella from the manufacturing of pulping machinery to the manufacturing of paper. This capability broadening required building of a paper mill whose history was examined in Chapter 5 of this volume. An essential detail here is that the paper machines for the new mill were supplied by a Swedish manufacturer, but the machinery for the mechanical pulp mill and the steam power plant were delivered by Tampella.

The third episode in Tampella's expansion from forest industry machines to their end-product businesses started during the Second World War. This process was not as revolutionary as the previous two because it actually implied only the replication of the first using a different raw material, namely recycled paper.[4] Therefore only minor modifications to the existing technologies were needed.[5] The fourth episode was similar to the third in the sense that it also was based on the deepening of Tampella's capabilities in mechanical pulping. In this case, instead of recycled pulp, Tampella developed a new semi-chemical pulping method. A plant using this technology to produce pulp used as an ingredient in various cartonboard grades was built in 1952. Although the plant did not create a major new business, it did represent an important technological step for Tampella because it could be used as a testing facility for Tampella's new Fluting mill built in the early 1960s. Fluting is the material used in corrugated cardboard. Both the pulping machinery and the fluting machine were supplied by Tampella. As a consequence, Tampella simultaneously widened its range of forest industry products and forest industry machinery businesses. This process was soon continued as Tampella also built a corrugated cardboard factory in which part of the new Fluting mill's production could be converted into cartonboard packaging.[6]

The fifth episode implied broadening of Tampella's capabilities in the production of forest industry machines by adding large scale board machines to the previous pulping machinery range. This was permitted by a combination of Tampella's recently developed capabilities in pulping (the semi-chemical pulping mill) and board machinery manufacturing (the fluting mill). These capabilities made it possible for Tampella to expand its forest industry businesses by building a new board machine and a mechanical pulp mill based on recycled fibre.[7] Because one of Tampella's first board machines was delivered to its own mills, the company simultaneously became a large scale producer of recycled board.

The sixth episode in Tampella's long-term expansion took place in the latter half of the 1970s. At that time Tampella had invented a new mechanical pulping method, namely 'pressure groundwood' (PGW). In 1979 a pilot plant using this new technology was built next to the Tampella paper mill. In the following year Tampella decided to build a mechanical pulp mill (PGW) which remains the world's largest facility using that technology. The mechanical pulp produced was further refined into newsprint and board with a major new paper machine and a rebuilt board machine which started up in 1983. Tampella thus deepened its capabilities in mechanical pulping machines, which again resulted in expansion into the forest industry businesses. The last episode in this process started in 1987 when Tampella again broadened its capabilities in the making of paper and board machinery. This time Tampella started to produce paper machines, whereas it had previously built only board machines. At the same time the company decided to enlarge its development facilities by building its first pilot paper machine. Tampella's engineering works, now called the Pulp and Paper Machinery Division, received its first paper machine order the same year. Its second delivery was a paper machine to its own paper mill. This project, which has been the empirical focus of this book, was also Tampella's first reference for a machine producing coated papers.

To summarise, the analysis of Tampella's successive major technological change projects showed how its capabilities in the making of forest industry machinery were used to create new forest industry businesses within the same company. The analysis also shows how Tampella as a forest industry company has developed through consecutive technological discontinuities. A common feature of all of these projects presented above as seven critical episodes is therefore the connection between these two lines of business and the fact that most of the new technologies were self-developed or at least self-built. The coherence of the identified pattern suggests that it is not merely a way in which Tampella happened to develop its businesses. Instead, it can be argued that this mode of action had its roots in the business structure of the company. For example, it is even possible to maintain that the development of Tampella's different business ranges has been largely determined by their technological distance from the core business of the corporation, that is mechanical engineering.

This distinctive form of expansion seems to have both advantages and disadvantages for the corporate competitive position. On the positive side, building and maintaining its own forest industry facilities yielded significant turnover for Tampella's machine making business. It also provided one solution to the generic product development problems of this business. For example, out of the three possible ways for a machine producer to develop its products – its own reference facilities, pilot machines and co-operation with other producers – Tampella consistently applied the first. Reference plants of its own also created the credibility required for further orders. On the negative side, however, it can be argued that this approach isolated Tampella from the market to some extent and resulted in a disintegrated business structure in both machine making and forest industry. For example, Tampella developed several unique paper machine constructions which were not reproduced in further deliveries. The uniqueness of the machines also gave rise to idiosyncratic products which later remained the property of Tampella's forest industry facilities. Moreover, in the long run Tampella's business portfolio also spread into numerous businesses which faced largely different kinds of market conditions.

The case of Tampella suggests two general reasons why this form of technological development may, despite its inherent problems, emerge. First, this is because the differences between the two types of business activities may actually encourage their location within a single company. It is argued here that the differences between paper machine and paper products businesses (see Table 7.2) may form such a mutually complementing combination. Second, a long tradition of systematic development of the technological link between machine building and paper production is inevitably difficult to break. Similar continuities have been found in recent studies that have noted that behavioural patterns created in the founding phase of a company probably remain (Boeker 1989). This is partly because they invalidate the effect of negative feedback (cf. Levinthal and Myatt 1994). In this case, the co-operative link between machine making and forest industry businesses was also enhanced by the fact that Tampella built most of the new facilities itself.

However, demonstration of this long-term pattern in Tampella's technological development has not been motivated only to provide these conclusions. In fact, this process has been followed from the point of view of the Tampella's forest industry development without systematic references to the ways in which the company has expanded its other businesses. This is simply because expansion into forest industry products best reveals the interrelationship between the development of machine production and Tampella's other businesses. Most importantly, it offers a clue in the search for a contextual explanation of the technological discontinuity in focus. The analysis above shows that the discontinuity analysed in detail in the preceding chapters of this book is only one link in a chain of successive major technological change projects in which Tampella entered the ultimate markets for its machines. However, such projects are impossible to perform if sufficient material resources are not available. The next section therefore analyses the changes in Tampella's resource base.

Changes in the corporate resource base

In this section we describe how the amount of financial resources available to Tampella suddenly increased in the late 1980s in a way that encouraged managerial actors to initiate and implement major technological change projects. Before analysing the background of these temporary improvements in Tampella's financial position we first analyse the reasons for the company's long-term lack of financial resources.

Origins of Tampella's resource scarcity

The main reason why Tampella suffered from a lack of resources from the early 1970s to the late 1980s was the fact that most of the endeavours mentioned above had fallen short of expectations. We have already mentioned that Tampella's new newsprint machine (PM 3) did not prove successful in the early 1980s. In addition, the Fluting mill developed by the company had faced unexpected increases in its raw material expenses and Tampella's cartonboard machine had suffered from difficulties in the production and marketing of its products based on recycled pulp. As a consequence, only large divestments and

Table 7.2 Comparison of the main characteristics of paper machine and paper products businesses

Characteristic	Paper machine business	Paper products business
Cash flow	Occasional	Steady
Nature of products	Largely genuine	Standardised
Number of business transactions	Low	High
Key capability	Innovativeness in the creation of technologies	Efficient production of standard products

intensive cost control saved Tampella from bankruptcy in the 1970s. However, although Tampella survived it went deeply into debt. Still more debt was incurred from borrowing to finance the investments of the early 1980s. As a result, Tampella's debt per turnover ratio increased from 0.9 to 1.5 (see Table 4.2 previous). In addition to its debts, Tampella also had owners who did not want to increase Tampella's share capital. As an indication of this, the company's share capital was raised between 1971 and 1985 from only US$15 to 25 million, whereas the firm's capital expenditures between 1981 and 1983 alone amounted to US$550 million.

Tampella was thus in a vicious circle (cf. Masuch 1985). In brief, this means that as a result of long-term resource scarcity only a few major technological change projects could be implemented. Moreover, the experienced managers needed to execute them were not available and new managers were not motivated to join Tampella and conduct the projects needed to attract more people for later projects. The vicious circle behind Tampella's resource scarcity is illustrated in Figure 7.2. In situations where technological discontinuities were being initiated, the lack of financial resources may also have encouraged managerial actors to choose less risky, although equally less innovative technologies. For example, Tampella's financial state would not have allowed borrowing to finance a more ambitious technological discontinuity than what was performed in the early 1980s. In a way, Tampella was doomed to newsprint because entering the production of coated paper grades would have cost an additional US$100 million. This amount would have been more than 20 per cent of Tampella's investments between 1981 and 1983 and more than 30 per cent of the total budget for the whole project (Laurila 1997a: 231).

In summary, in the latter half of the 1980s Tampella's businesses were doing badly. The company was also deeply in debt. However, any major changes in the profitability of its businesses necessitated significant capital outlays. The next sub-section examines the contextual features which led to the breaking of this vicious circle in a short time.

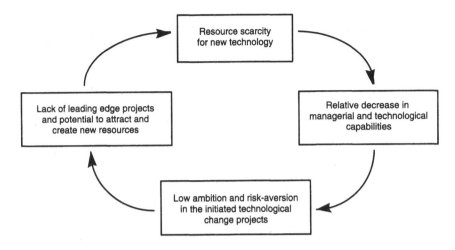

Figure 7.2 The vicious circle behind Tampella's resource scarcity

Moving from a vicious to a virtuous circle

We have already described how Tampella's new owners announced their willingness to finance major new investments in the company. We have not, however, analysed the background of this willingness and how significantly Tampella's financial position actually changed in those years. The greater willingness of the new owners to make capital expenditures was reflected by the fact that after the take-over in 1987, Tampella raised its share capital from US$34 to 43 million and in the following year to as much as US$65 million. Moreover, Tampella's share capital was still raised annually throughout the rest of the 1980s (see Table 4.2). At the same time Tampella's financing costs decreased from 10 per cent of turnover in 1986 to 3 per cent in 1988. The changes implied that investments were financed not only by borrowing but also by raising share capital, which enabled major new capital outlays. For example, Tampella's total investments between 1980 and 1987 were less than US$500 million whereas they exceeded US$1 billion between 1988 and 1990 (see Table 4.2).

The changes in Tampella's financial position were not, however, only the result of the new owners' will to help the company. Instead, there were two factors that made the new owners' task rather easy. The first of these was the financial deregulation of the Finnish economy that began in the middle of the 1980s after almost sixty years of tight governmental control. The banking sector was allowed to operate on the international finance markets, which resulted in a rapid increase in the amount of foreign capital in the Finnish economy. The Finnish economy became very liquid, and this in turn led to a simultaneous and mutually accelerating expansion of credit and increases in the value of real properties. Consequently, the quoted value of Tampella's shares on the Finnish Stock

Exchange started to rise in 1985 and before the end of 1987 their value had increased eightfold.

Another factor helping the new owners was that the financial deregulation made the market value of the Finnish industrial companies deviate notably from their quoted values. In other words, industrial companies were more valuable than they looked because their book value had not been adjusted to changes in the market. For example, a production facility might be sold at a price one hundred times above its book value. Therefore, corporations like Tampella whose profitability had been poor, but which also possessed significant property outside its business operations, became popular objects of speculation. In addition, because Tampella's share capital was small in relation to its business activities, a take-over would be relatively easy. Most importantly, however, because of the undervaluation of the properties of Finnish corporations, opportunities for increased borrowing appeared. This is to say that the rapid increases in the market value of this property made previous debt look relatively smaller. The increase in the quoted value of shares also permitted the companies to increase their share capital and to improve their solidity.

In summary, the resource scarcity that had plagued Tampella for many years was temporarily overcome in the late 1980s and no longer operated as a restriction.[8] This also means that as far as material resources are concerned, Tampella was suddenly offered an opportunity to move from the previous vicious circle into a virtuous one. The main characteristics of this virtuous circle are illustrated in Figure 7.3. A temporary resource surplus encouraged managerial actors to initiate ambitious technological change projects. Such projects would then accelerate managerial turnover, resulting in new managerial capacity with the experience and capabilities needed to formulate new competitive projects. Such projects would also have the opportunity to increase the available material resources for long-term technological development. As a consequence of the increase in material resources, we can therefore expect at least temporary changes in the managerial actors' mode of operation. For example, if Tampella's financial resources in the late 1980s had been scarce, the ambitiousness of the new paper machine (PM 2) could have been perceived as a threat. However, because material resources at that time were abundant it was in fact perceived as an opportunity. The technological risks inherent in the technological discontinuity were not perceived as hazardous by the actors, because the resources available led management to consider them irrelevant. Various kinds of technological discontinuities may therefore represent an ordinary and almost obvious course of action from the perspective of the managerial actors initiating them.

However, material resources alone cannot explain the emergence of technological discontinuities. This is because, in addition to resources, managerial actors also need models or ideas concerning how to use those resources. Such models give the discontinuities their substance: what is actually being done and how. For example, one such model is the traditional pattern in Tampella's technological change projects; managers initiate projects in which the new

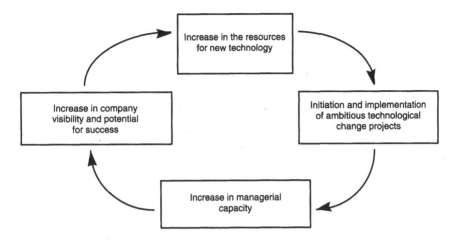

Figure 7.3 The virtuous circle of corporate resource development

technology can be supplied by the company itself. However, even a model of this kind does not specify into which forest industry business Tampella should invest and when. It cannot explain why the technological discontinuity occurred at the Tampella paper mill after several decades of relative decline and at a time when other Tampella forest industry businesses also lacked new technology. The next section therefore examines Tampella's competitive environment where, alongside the positive changes in Tampella's corporate resources, the company was also becoming free to expand into paper machine manufacturing and Tampella's competitors were moving into new paper products and production technology.

Changes in the corporate business environment

This subsection aims to clarify why the technological discontinuity which renewed Tampella's paper products and production technology was so long delayed, but was then abruptly carried out in the late 1980s. To understand why Tampella's traditional pattern of technological development was not applied to its paper mill until half a century had elapsed, we first take a look at the structure of Finnish paper machine manufacturing. The analysis of how the Finnish paper machine market and Tampella's role in it have developed starts with the origins of that industry in Finland and then moves to the changing patterns of cooperation between specific firms in the sector.

The rise and fall of co-operation between Finnish paper machine manufacturers

The forms of co-operation between the Finnish paper machine manufacturers have significant impact on the long-term development of Tampella's paper industry operations. As mentioned previously, Tampella had manufactured pulping machinery from the late nineteenth century onwards and also board machines after the Second World War. This extension of production was feasible at that time because Tampella had to convert its previous gun and locomotive factories to a new line of production. This conversion was part of the overall development among the Finnish paper machine manufacturers, which now started to supply the paper and board machines which had previously been imported to Finnish mills.[9]

Tampella was thus heading towards the paper machine market after being a paper producer since the 1930s. However, there were also three other Finnish companies and engineering works operating in this field. The important fact for the analysis here is that three of them, that is Wärtsilä, Valmet and Tampella, soon began to co-operate with each other. The background of this co-operation (see Laurila 1995: 119–129) was the fact that all the Finnish paper machine manufacturers faced similar problems. One of these was that in many of the early competitive biddings the three lowest bids were made by Finnish suppliers, which further reduced their offers by around 10 per cent before agreement was reached. This loss of margins restricted their opportunities for product development. Moreover, they all needed orders, though they lacked references from earlier deliveries.[10] Thus the Finnish manufacturers had only four alternatives. First, they could restrict themselves to a specific market such as Russia and Eastern Europe. Second, they could develop new innovative machine designs jointly. Third, they could reduce the prices of their products and fourth, they could try to deliver their first references to a friendly customer. The Finnish paper machine manufacturers used all of these alternatives and as a result, their position on the international paper machine market started to improve.

Co-operation with other producers enabled Tampella to make forest industry machines a prominent business. This development was further enhanced in 1969 when the co-operation between the Finnish paper machine manufacturers was placed on a formal basis through the 'TVW' agreement by Tampella, Valmet and Wärtsilä. The content of this agreement conformed with the principles developed earlier. Tampella was to specialise in board machines, Valmet in paper machines and Wärtsilä in machinery used in paper finishing. The TVW co-operation comprised co-ordination of research and product development, production and marketing of paper and board machines by these companies. The biggest difference compared with the earlier arrangements was that co-operation between these manufacturers, which at this time already represented one-fourth of the international market, was now extended to all market areas.

The most relevant implication of the TVW agreement for Tampella's paper industry activities was that Tampella was not allowed to produce paper machines. For example, in line with the agreement, Tampella's new paper machine (PM 3) in the early 1980s was supplied by Valmet. Although the co-operation somewhat restricted the ways Tampella could extend its businesses, the company also benefited, especially because until the late 1960s it was relatively weak in machine construction compared with the other partners.[11] Tampella was also more dependent on the co-operation than the other partners, which had the resources to handle many of their market activities outside TVW.

Co-operation between the TVW partners was important mainly because it directed Tampella into board machine construction. However, various contradictions soon eroded this co-operation. In fact, cracks in the TVW facade were already apparent in the 1970s. This was because both the partners' intentions to expand their operations into paper machine manufacturing and their available resources for this expansion varied. For example, paper machines had become the most important business for Valmet whereas Wärtsilä had concentrated on shipbuilding. Tampella was still the smallest machine manufacturer of the three. Co-operation was also challenged by technological development in paper machine manufacturing. For example, the prospects of Wärtsilä as a supplier specialising in paper finishing facilities were threatened by technology development which permitted integration of paper finishing with the rest of the paper machine. Moreover, the markets for board machines had contracted in the early 1980s and thus Tampella also had interests in extending its previous range of operation.

Tampella became free to expand its machine making activities in 1986, when co-operation between the Finnish paper machine manufacturers formally ended. Before this both Valmet and Wärtsilä expanded their operations through international acquisitions which did not conform to the previously defined boundaries. For example, Valmet entered the manufacturing of board machines which was previously Tampella's territory.[12] The final blow to the TVW co-operation was the merger of Wärtsilä's paper machinery production with Valmet. Moreover, by the end of 1987 Valmet also bought up all the other Finnish paper machine producers and also offered to buy Tampella's board machine production. After these acquisitions Valmet's product range in paper and board machinery was almost complete, whereas Tampella was left with only its references in board machines. Moreover, as the TVW co-operation ended, Tampella lost its previous channels to international markets and its opportunities for product development were minor compared with those of its competitors, for example Valmet.[13] In this situation Tampella had only two alternatives. It would either have to sell its paper machine production or compete. Tampella had many reasons to give up but instead in February 1987, about a month before it was taken over, it chose to extend its product portfolio in machine manufacturing to include paper machines.

To summarise, the development of Tampella's activities in machine making were affected by changes in its business environment and especially by its interaction with the other Finnish paper machine manufacturers. The analysis also shows that the arrangements Tampella made in the paper machine market put its different forest industry businesses into an unequal position. In particular, because of Tampella's specialisation in board machinery the company's only paper mill received less attention than the other forest industry units. Co-operation between the Finnish paper machine manufacturers thus provides a partial explanation for the late timing of capital expenditures at the Tampella paper mill. On the other hand, the break-up of the same co-operation also had a dramatic impact on the Tampella paper mill. Expansion by its competitors at the same time allowed Tampella to extend its operations into paper machine production and also created strong pressure to do so, as it threatened the very existence of Tampella in the business. In fact, Tampella's goal of moving into the paper machine market made its paper mill an important supporter of this process, as the company was also able to continue its traditional approach to expansion through successive major technological change projects in this field of the forest industry. However, the issues taken up so far cannot explain the ambitiousness of the technological discontinuity at Tampella paper mill. In other words, Tampella would not have needed to enter a technically inconvenient and (from a market perspective) relatively small product area. To understand the motivation to this we briefly describe what was happening among the Finnish paper producers at that time.

The shift from newsprint to coated paper grades in the Finnish paper industry

The details presented above showed that Tampella had reason to initiate the technological discontinuity at its paper mill in the late 1980s. Tampella clearly needed references on paper machine deliveries and its paper mill was thus a suitable new customer. However, to understand why Tampella did not continue as a newsprint producer, but instead chose to enter the manufacturing of coated paper grades, we briefly summarise what had happened within the Finnish paper industry during a period of a few years preceding Tampella's discontinuity.

Tampella's move into coated papers was motivated mainly by the fact that since the early 1980s – due to the virtual disappearance of the margin between its price and costs of production – newsprint had not been a popular product with Finnish paper producers. The relative decline of this line of production was illustrated previously in Figure 5.1. This does not mean that all of the Finnish paper industry companies gave up their production. However, many did and in particular, no company after Tampella in 1980 initiated expansion of newsprint production with virgin fibre. The overall shift from newsprint to coated papers is also indicated by the fact that in the early 1980s there were thirteen paper mills in Finland operating in this field. None of these built a new newsprint machine in 1984–1988, although there were four of them between 1977 and 1983. In contrast, between 1977 and 1988 the same mills built seven new paper machines for producing coated paper grades, five of which started up between

1984 and 1988. Thus between the early and late 1980s newsprint became unpopular among those firms operating within the same product area with roughly the same production costs as Tampella. Moreover, several older newsprint machines were rebuilt to produce calendered and coated grades during the 1980s.

Coated paper grades were thus fashionable at the time when Tampella's technological discontinuity was initiated. It can therefore be argued that Tampella was conforming to the trend among Finnish paper-producing firms. Simultaneously, by supplying technology to such a facility Tampella was also developing capabilities for the machine construction business. Nevertheless, the paper grade (MFC) adopted cannot be considered an obvious choice at the time. In fact, it would not have been unusual to choose the technologically more conventional coated (LWC/off-line) grades (see Figure 5.2). However, although MFC grade was technologically demanding, some of Tampella's competitors had already entered the field through similar kinds of projects. Their concept could therefore be imitated and as a consequence, Tampella would obtain an important reference for a demanding type of coated paper machine.

Conclusions: the emergence of technological discontinuities as a situationally conditioned managerial activity

We described above how Tampella developed its businesses through successive technological change projects and how the situational factors related to both corporate resources and corporate business environment embodied this development in the form of the technological discontinuity at the Tampella paper mill. Thus situational factors permitted Tampella's long-term pattern of technological development to continue as obstacles were now momentarily removed.

The situational factors did not, however, operate without reference to human action. In fact, they produced the conditions which permitted the managerial actors to initiate and implement the discontinuity. The traditional ways of developing technology, the available resources and the characteristics of the corporate business environment are interpreted by the actors when they consider future actions. This means we can expect technological discontinuities to emerge when there are several factors simultaneously encouraging managerial actors to undertake them. In the case of Tampella, the abrupt increase in financial resources contributed to the ambitious discontinuity by permitting the acquisition of state-of-the-art technology. This further permitted mobilisation of new managerial resources, thus increasing the chances for successful new technology implementation. On the other hand, the managerial actors at Tampella became bolder because they felt the discontinuity both fashionable and not too risky when compared with the corporate resource base. In other words, technological risks connected with the shift from newsprint production to high value-added magazine paper grades were easier to accept when changes in business environment and material resources appeared to back up the discontinuity. This also means that the changes in these situational factors were perceived by the managerial actors as opportunities to promote their divergent objectives and goals.

The material presented in this chapter permits us to take yet another view of mobilisation of managerial actors in Tampella. As mentioned above, mobilisation in this study has referred to temporary coalitions of actors, each comprising several individual managers. However, another way to use the mobilisation conceptualisation would be to consider mobilisation as a cluster of long-lasting coalitions in management with their own distinctive ideologies, means and objectives. A socio-historical approach of this kind to the case of Tampella would reveal several competing mobilisations in management. In Table 7.3 we have identified three such mobilisations as coalitions of dominant managerial actors at each stage of Tampella's technological discontinuity. The implication here is that each contextual condition determines the kinds of coalitions of managerial actors that become dominant.

However, there is yet another way to develop the conceptualisation of management mobilisation. For example, the divisions between locals and cosmopolitans and between corporate and mill level can be used as the basis for identification of the competing mobilisations at Tampella. The resulting four types of actors are illustrated in Figure 7.4. To briefly characterise the different mobilisations created by these two dichotomies, the previously dominant corporate level mobilisation was oriented to improving Tampella's stance, mainly through institutional relations. For example, it aimed to secure Tampella's position through different kinds of institutional alliances (e.g. the TVW agreement), thereby somewhat neglecting independent development of the various businesses. In contrast, the cosmopolitan mobilisation at corporate level, which was taking over in the late 1980s, emphasised professional corporate and business management. From this perspective, Tampella was to manage in competition with others by integrating market demand and technological advancement in all of its business units. On the mill level, the local mobilisation emphasised incremental development and 'eaking out a living'. This tradition had been created during the long-term resource scarcity the mill had suffered before the late 1980s. In contrast, the cosmopolitan mobilisation at the mill level aimed at technological superiority through adoption of new technology.

Competing mobilisations in management therefore partly explain what kinds of technological discontinuities are initiated or how they are implemented during the period of their domination. Moreover, it is always possible to identify competing mobilisations located on different levels of the managerial hierarchy and even across separate organisations. In specific situations the conflicts between these mobilisations may produce changes in the managerial power structure. In addition, the specific circumstances in which these competing mobilisations collide determine which mobilisation must yield.

Table 7.3 Contextual and intentional factors contributing to Tampella's technological discontinuity

	Contextual factors			Intentional factors: dominant managerial actors in each period
	Amount of resources	Competitive situation	Model of technological development	Actor intentional/dominant actors
Before 1987	Scarce	Partly restricted (co-operation in machine manufacturing)	Simultaneous expansion of machine making and forest industry activities	Irresolute owners Technology enthusiasts Production functionalists
1987–1989	Abundant	Open	Simultaneous expansion of machine making and forest industry activities	Risk-seeking owners Technology enthusiasts Business innovators
1989–1991	Decreasing	Open	Simultaneous expansion of machine making and forest industry activities	Risk-seeking owners Experienced generalists Technical specialists

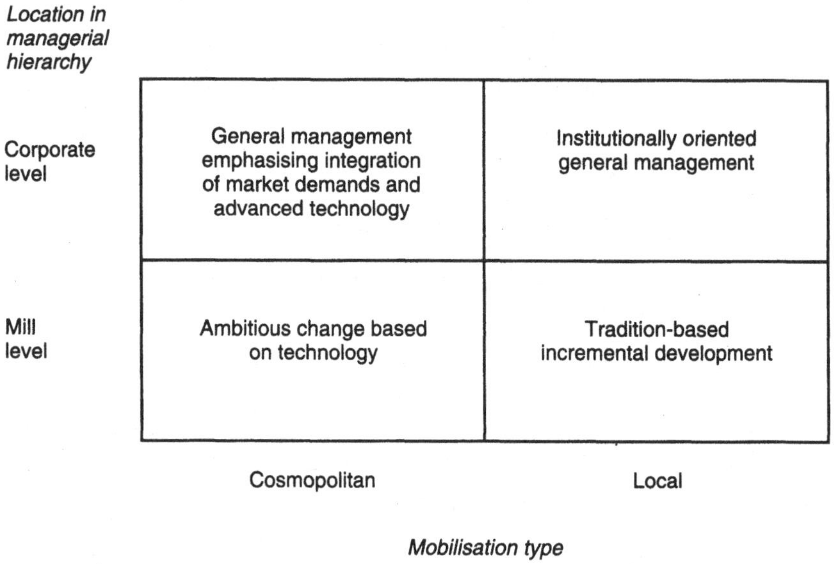

Location in
managerial
hierarchy

Corporate level	General management emphasising integration of market demands and advanced technology	Institutionally oriented general management
Mill level	Ambitious change based on technology	Tradition-based incremental development

Cosmopolitan   Local

Mobilisation type

Figure 7.4 Types of management mobilisation at Tampella

8   Further challenges in the study of managing technological discontinuities

This chapter brings together the main themes of the book and reflects on the theoretical and practical implications of the material presented. In the first section of the chapter we summarise the findings on the interrelationships between technological discontinuities and the mobilisation of managerial actors and elaborate on them with reference to recent literature on the field. In the second section we move to examine potential developments in the conceptualisation of management mobilisation. In the third section we reflect on the lessons of this book for future study of technological discontinuities. Finally, the concluding section of the chapter includes some implications of the study for managerial practice.

Discontinuities as issues of mobilisation

The aim of this study has been to demonstrate how technological discontinuities are produced by mobilisation of managerial actors. The study is therefore in line with the actor-focused stream of research within management studies which has demonstrated and elaborated on the effects of management on various organisational outcomes. One of the main findings of previous studies in this field is that the background and composition of managerial actors are reflected in the development of business firms. For example, demographic diversity within top management teams has an effect on management creativity and various characteristics of the decision making process (Bantel and Jackson 1989). These studies have also elaborated on the relationships between separate managerial actors and demonstrated the effects of their mutual interaction on various indicators. For example, a recent study by Westphal (1997) noted that corporate CEOs are capable of eliminating the effects of increased board independence on corporate strategy. Other recent studies have examined the formal sources of power and showed that the impact of actors involved in the management of corporations is greatly dependent on their will and skill to convert potential power into actual influence (Pettigrew 1992; Pettigrew and McNulty 1995).

However, the relevance of the actor-focused stream of research here is that in addition to demonstrating that management is important and consists of several actors, it has also shown that organisational change requires specific kinds of intra-organisational processes. Managerial actors have been found to interact more intensively at times of major changes in order to create shared grounds for action (Westley 1990; Liedtka and Rosenblum 1996). This is necessary to ensure efficient processing of information between different actors (Eisenhardt and Bourgeois 1988). Discontinuous change thus necessitates cognitive sense-making with which the actors try to overcome their uncertainty and confusion concerning the action to be taken. Moreover, previous research has also identified factors such as restricted pay dispersion among top management members (Hambrick and Siegel 1997) as critical for the emergence of intra-managerial collaboration in a situation of major discontinuity.

Although we can acknowledge progress in the conceptualisation of management mobilisation for discontinuous change, it can be argued that the mainstream of actor-focused management studies is far too mechanistic to make a rapid breakthrough in this field. In this study we have assumed that managerial actors have diverse motivations for their action. This assumption is in line with recent studies (e.g. Parker and Dent 1996) suggesting that managerial divisions are inevitable because of the distinct frames of meaning within organisations. A study by Dutton and Duncan (1987) can be used to demonstrate the difference between such an approach and the previously dominant forms of study. This study argues that social momentum for change (i.e. mobilisation of managerial actors) emerges as a consequence of a diagnosing process in which the different actors find opportunities for new ventures. Cognitive diagnosing on the specificities of the managerial task at hand is therefore a precondition of social mobilisation towards the chosen way of action. The strength of this study and some other related studies (Jackson and Dutton 1988; Dutton and Jackson 1987; Thomas and McDaniel 1990; Sharma 1997) is their suggestion that intra-organisational processes determine whether given external conditions are interpreted as threats or opportunities. However, we believe that this study omits various relevant aspects of management mobilisation because it recognises realisation of managerial objectives as the only aim of the actors.

In this study we therefore assumed that management is internally divided and that the divisions are diverse and constantly changing. Although these aspects are important, managerial actors are not formed only on the basis of such concrete factors as formal hierarchical divisions or demographic similarity but also on the basis of various informal and ambiguous factors. One of the reasons for the diversity within management is the often ignored fact that instead of being sovereign, managers are frequently weak in relation to the changes they are expected to effect (Teulings 1986). Managers are also often afraid of becoming redundant (e.g. Simpson 1997). On the other hand, managers' career aspirations and modes of action vary. For example, Leavy and Wilson (1994) have found four separate roles for individual CEOs. They propose that CEOs may act as builders, inheritors, revitalizers and turnarounders in heading their companies. As a consequence, we see that both the 'hard facts' of managerial work and the subjective individual and collective interpretations influence the creation of potentially conflicting managerial actors. In order to conceptualise such divisions we need to be sensitive to the formal hierarchy, informal social relations and diverse patterns of meaning within management.

One of the major sources of conflict within management is the substance of managerial work. Such starting points are in keeping with more general tendency to move away from conceptualising managerial actors in isolation from their organisational context (cf. Hartley *et al.* 1997; Leavy and Wilson 1994). More concretely, in this study we showed how the specificities of paper production technology divided managerial actors into conflicting groups. This is an example of managerial hierarchies being political, but instead of resulting in open conflicts, the study also showed that these political features of management are embedded in the divergent managerial objectives concerning the actual business operations.

Managerial actors are also products of different professional identities or specific experiences shared by each managerial cohort. These collective qualities supply managers with an intellectual tool kit to be used in specific problematic situations (Spender 1989). For example, the newcomers to the management of Tampella in the late 1980s tried to transform the company according to the current dominant recipe (cf. Porac and Baden-Fuller 1989; Hellgren and Melin 1992) in the Finnish paper industry. In contrast, some of the old-timers developed Tampella gradually, based on tradition and incremental extensions of the capabilities it already possessed. That is to say the conflicts were not only between individual managers or groups of managers, but also between different lines of thinking. For example, the main divider in the management of Tampella was the tendency to emphasise either the risk of bankruptcy or that of competition (Ghemawat 1993) related to technological discontinuities.

It seems that these conflicting management styles and preferences are also important because they function as 'lightning rods' for social tensions within management. In other words, managerial objectives can be used as a camouflage for the personal interests and goals of the actors. This study argues that these personal interests and their correspondence to the specificities of the technological discontinuity at hand are an important contributor to the mobilisation of managerial actors to support the discontinuity. Technological discontinuities thus tend to be coupled with management mobilisation which emerges largely out of the opportunities offered by the discontinuities to both implement the managerial objectives of the actors and to secure personal benefits such as career advancement, professional experience and reputation. In this study we have demonstrated how the various benefits connected to technological discontinuities may at least temporarily connect actors with clearly different backgrounds, objectives and goals.

The mentioned findings on management mobilisation are related to recent developments in management and organisation literature. For example, the argument that there is no one unambiguous basis for the mobilisation of managerial actors is in line with the general claim that social relations within management have to be continuously renegotiated (Strauss 1978). Other theorists have noted that specific organisational forms, such as management mobilisation, are vital at times of discontinuous change (e.g. D'Aveni 1994). This is because without such forms the social momentum needed to overcome the challenge of change cannot be created. A more elaborate recent study on these issues has found that discontinuous change requires collaborative leadership including constellations of actors playing distinct roles (Denis *et al.* 1996: 673). Such forms have been noted to require common beliefs which are created in joint experiences of the concrete change activities (e.g. Smith *et al.* 1994; Westphal and Zajac 1995). Such consensus has been found to emerge first in those parts of the managerial hierarchy where the benefits of the change are the most explicit (Markoczy 1996).

Thus there is relatively wide agreement on the importance of management mobilisation, although knowledge on these processes is insufficient. However, it is important to note that discontinuities are not caused solely by the different forms of management mobilisation. This is, first, because mobilisation is a temporary and fragile phenomenon. Mobilisation does not indicate a unified management or corporate culture as even one critical shortcoming may lead to disintegration of the alliances and coalitions. This is in line with arguments according to which the tendency of managerial actors to interpret the discontinuities either as an opportunity or as a threat makes the mobilisation of managerial actors susceptible to rapid decline (Soeters 1986). Mobilisation is also fragile because managerial actors may have joint beliefs which do not necessarily create solid commitment among them (Denis *et al.* 1996; Markoczy 1996).

An even more important reason to restrict the power of management mobilisation in the production of technological discontinuities is that it is in itself a contextually conditioned phenomenon. By systematically examining the context in which the technological discontinuity at Tampella was initiated and implemented we have aimed to show that management mobilisation actually only mediates the effect of environmental conditions. This is to say mobilisation utilises the opportunities created by temporary changes in its context. Thus highlighting mobilisation has not implied a choice between voluntaristic and environmental sources of technological discontinuities. Instead, it is assumed that these two sets of factors are mutually interdependent in the sense that contextual factors continuously produce situations whose specificities can be used by the actors to promote their own intentions (Knorr-Cetina 1981). The aim has thus been to avoid the dominance of either the strategic choice or environmental determinism type of approach (see e.g. Astley and Van de Ven 1983; Hrebiniak and Joyce 1985) by taking account of the interdependency of these contextual and intentional factors in a historically specified situation. Using the concepts of Wilson, Hickson and Miller (1997), we want to strike a balance between the contextualist (which proposes that context is pre-eminent over management) and the individualist (which proposes that managerial action is fundamental for organisational change) arguments in our approach.

The findings of the study which showed that simultaneous changes in Tampella's competitive position and corporate material resources allowed managerial actors to initiate and implement a technological discontinuity are in keeping with some recent works. First, the case of Tampella confirms that the demographic effects of top management teams are conditioned by environmental uncertainty (Carpenter 1997). More importantly, the findings of this study conform to a recent study by Allmendinger and Hackman (1996). They empirically demonstrate that organisational changes occur especially when the connection between organisational actions and organisational resources is strengthened simultaneously with increases in the operational autonomy of the individual organisations. Consequently, it can be argued that contextual factors favour one mode of action over another. For example, on the basis of the case presented, we contend that in situations of resource abundance the actors tend to choose ambitious discontinuities over minor improvements, because the former create the greatest expectations of individual and collective benefits for the actors taking part. This happens even though the success of a discontinuity can never be anticipated beforehand. Various concrete problems related to the discontinuity must also be resolved even though a single, obviously right course of action cannot be found. In the paper industry, for example, major technological change projects have to be formulated years before these projects are actually completed. At that time it is impossible to know for certain how profitable or successful the implementation of the specific technologies will turn out to be.

Because mobilisation of managerial actors to initiate and implement technological discontinuities is contextually conditioned we may expect to find such forms in firms which for example are suddenly supplied with new material resources. The most interesting cases of this kind, however, are those in which relatively limited additional resources lead to major changes in technology (cf. Hamel and Prahalad 1994). In this spirit, we have examined Tampella, which after a long period of resource scarcity was given substantial resources for a few years. We acknowledge that from an external point of view, Tampella's discontinuity is not the most unexpected phenomenon. The discontinuity was also conventional from the point of view of the Finnish paper industry. It would actually be surprising if major increases in financial resources did not encourage managers to ambitious action. Nevertheless, Tampella is relevant largely because although it had resources for a while, it also had an unconventionally weak experience base from which to start its leap towards the leading edge of its range of business. As a result, by examining Tampella this book has provided insights into the process through which competitive technologies may be created even from relatively weak starting points.

Evolution of mobilisation conceptualisation

In this book we have constructed a specific kind of conceptualisation for the mobilisation of managerial actors. We do not, however, see this conceptualisation as the final one. On the contrary, we believe that many new conceptualisations are about to emerge as a result of further studies. In this subsection we discuss the limits of the conceptualisation presented here and anticipate the characteristics of the forthcoming, more advanced conceptualisations.

Developments within the previous field of study

We can identify several directions in which the previous conceptualisations on management mobilisation deserve to be developed. First, we see that current research has only made half of the journey from rationalistic to subjectivistic conceptualisations on management mobilisation. For example, in this study we inductively constructed managerial actors as idealised patterns of the thinking and acting represented by each actor. This was necessary to make management mobilisation understandable as temporary forms of alliances and coalitions between these actors. The approach adopted can be considered innovative in the sense that previously management literature has largely neglected the specificities of management as an occupation and ignored the sources for the motivation of the individual managers (Van Maanen and Barley 1984). Consequently, management has traditionally been conceptualised mainly through its general objectives and functions (Willmott 1987). In Anthony's (1986) words, the emphasis has been on the official theory at the cost of the real one. Nevertheless, we acknowledge that the view of managerial actors proposed in this book may eventually prove simplistic. This is because it can be argued that managerial actors are not fixed to specific objectives and goals and they may promote different interests in successive situations and even several interests in the same situation. As a consequence, if managerial actors are presumed to promote such a variety of interests at the same time, the possibilities for forming coalitions and alliances in management are multiplied.

Thus there is a clear connection between conceptualisations on management mobilisation and ontological assumptions on managerial actors. We accept the views that social interaction within management, or for that matter between any social entities, is ambiguous. In other words, we believe that in many ways managerial action remains a complex and informal activity whose content and meanings are difficult to capture with any conventional research intervention. As a consequence, research on management mobilisation should focus both on conceptualisation of managerial actors and on the processes through which these actors are constructed. Nevertheless, we do not want to label mobilisation of managerial actors ambiguous only because the phenomenon is difficult to investigate. Instead, we believe that our understanding of these processes may still significantly change through further empirical study.

Further study of management mobilisation should be developed with two kinds of empirical research. First, further development of the finding that mobilisation involves temporary coalitions of managerial actors requires in-depth studies in specified corporate contexts. For example, the claim that instead of having relatively constant preferences and aspirations, managerial actors may actively switch their preferences (Grieco and Lilja 1996) is difficult to demonstrate empirically. Ethnographic research methods (e.g. Kunda 1992; Watson 1994), that is close involvement with the managerial community under study, seems one prerequisite for further progress in the conceptualisation of management as a temporary coalition of coalitions. Second, in addition to going more deeply into the research on managerial communities, various cross-sectional designs are needed for study of these issues. This is especially because in order to assess the genuineness of the forms of mobilisation constructed in intensive studies we need comparative work in which the same issues are studied in deviating settings. Moreover, because it seems warranted to believe that mobilisation of managerial actors is contextually determined, we should do research on a variety of contexts. In part, such research would enable making distinctions between particular managerial practices and general mechanisms of mobilisation. In a sense, turning the present conceptualisations on management mobilisation into a coherent body of literature will require much research in various national and business settings. This kind of reasoning conforms to the research strategy adopted for example in the recent research programme on national business systems (Whitley 1992; Whitley and Kristensen 1996).

Regardless of the methodology adopted in further studies, we believe that conceptualisations of management mobilisation will benefit from the use of a relatively concrete point of reference to which the often ambiguous forms of managerial action can be related. For example, in this study the forms of technological change and the ways in which different managerial actors have contributed to them have permitted us to identify distinctive features for each actor. We believe that such concrete sources of comparison are needed in the further development of those three dimensions on which management mobilisations differ from each other. These dimensions were already emerging in this study but need to be further elaborated in the forthcoming studies. First, duration of mobilisation refers to the variance in the amount of time between mobilisation emergence and disintegration. Second, comprehensiveness of mobilisation refers to the variance in the relative proportion of managerial actors committed to the activities promoted. Third, intensiveness of mobilisation refers to the variety in the amount of overlap between the objectives and goals of the actors involved.

Extending the scope of study

We have analysed the methodologies and conceptual improvements in the study of management mobilisation as it was defined here. We have therefore thus reflected on how to study these issues with more sophistication and anticipated the forthcoming findings as a result of such work. However, interesting developments can also be found by redefining the current object of study. Most concretely, this means that instead of examining how management mobilisation emerges and how it influences various organisational outcomes in individual companies, we could also examine mobilisation as a phenomenon exceeding definite organisational boundaries within a sector or a group of companies. One obvious reason for this is the fact that business firms have become increasingly permeable and relate with co-operative instead of competitive forms of behaviour to their competitors. It seems therefore that there is sufficient reason to observe how mobilisation of managerial actors in one company is related to mobilisations in other companies.

The contextual firm-in-sector perspective adopted in this study revealed similarities between technological discontinuities in different paper industry companies. We also noted that opportunities to imitate specific technological constructions encouraged initiation of an ambitious change project. These processes were not, however, systematically investigated here and therefore the role of knowledge transfer between different companies (see e.g. Hamel 1991) in the mobilisation processes needs to be elaborated in further studies. Interestingly, some recent work has emphasised that companies tend to imitate rivals which are relatively similar but which have also developed substantially more advanced capabilities in a specific field (Lane and Lubatkin 1997). In research covering processes between companies it is important to note that production technologies will never be as abstract as forms of managerial hierarchies. Technologies therefore are not as susceptible to organisational isomorphism (DiMaggio and Powell 1983). Nevertheless, we acknowledge that because production technologies for example in the paper industry have long life-cycles, imitation of others is really one way for the managerial actors to manage the uncertainties and risks related to new technology.

Executive migration is another issue calling for the inclusion of several companies in the study of management mobilisation. In this study we showed that movement of managers across organisations is necessary to provide new managerial resources for a company implementing an ambitious technological change project. We also noted that the incoming managers greatly influence the process of mobilisation by further increasing the diversity of the managerial actors involved. Previous studies (e.g. Child and Smith 1987; Boeker 1997) have demonstrated that newcomers with their previous experiences are an especially important source of corporate change. Nevertheless, for example the relationship between management migration and mobilisation processes has not been systematically explored. Additionally, we believe that studies of management mobilisation should be extended to cover actors other than managers. It has already been noted that management consultants are influential in initiating discontinuous change although they have only limited opportunities to effect successful implementation (Ginsberg and Abrahamson 1991).

We believe that inter-disciplinary work is needed for further developments of research on management mobilisation. Such work is critical for clarification of how individual managers form collectivities and how a variety of social factors is realised in the acts of individual managers. Moreover, it seems that management mobilisation should be considered more as a joint and also partly unintentional phenomenon. However, we believe that social movement theory is one credible source of such conceptual development.[1] One strength of this literature is also that it crosses many of the disciplinary boundaries which characterise previous management and organisation literature. In building the conceptual framework of this study we noted that social movement theory has already been applied to the study of management mobilisation and discontinuous change (e.g. Zald and Berger 1978; Soeters 1986; Davis and Thompson 1994). Nevertheless, we argue that social movement theory has not been fully utilised in management studies in general and in the study of management mobilisation in particular.

One general factor to enable further applications is the fact that social movement theory has recently progressed from metaphorical and organisation theoretical streams in a more actor-focused direction. In other words, concepts and insights emphasising subjective experiences of actors have been supplementing previous rationalistic tendencies (see e.g. Klandermans and Tarrow 1988; Taylor 1995; Scott 1998: 178). These new tendencies make it easier to utilise the substantial similarities between social movements and management mobilisation in further comparative work. Social movements, as well as management when mobilised to initiate and implement discontinuous change, are collectivities characterised by visionary leadership (e.g. metaphorical use of language), abandonment of conventional modes of action (e.g. working long hours) and construction of enemies (the acts of competitors). Although the similarities between social movements and management mobilisation thus seem clear, they have not yet been thoroughly elaborated. Moreover, so far the collective and emotional aspects of management have been treated in several relatively unrelated research traditions and without systematical extensions to previous literature on social movements.

Another reason to encourage further applications of social movement theory on this field is that in general, this body of literature has aimed to conceptualise collective action as a phenomenon emerging in its ongoing context. Such an approach is especially relevant for further extensions in the study of mobilisation of managerial actors beyond the boundaries of an individual company. We also believe that research on the relationship between executive migration and management mobilisation would benefit from recent developments in social movement theory. For example, current studies have found that social movements change mainly through the entry of new recruits (Whittier 1997).

Further study on technological discontinuities

In this chapter we have already explored some paths forward in the study of management mobilisation and discontinuous change. This section concentrates on progress in the study of technological discontinuities.

The present study suggests that the previous tendency to consider technological discontinuities a revolutionary, rarely occurring industry-level phenomenon (e.g. Tushman and Anderson 1986; Anderson and Tushman 1990) has taken attention away from the relatively frequent forms of technological discontinuities taking place at firm level. Although radical innovations making

previous technologies immediately obsolete were rare, technological discontinuities are both common and critical to the management of individual
companies. This is especially true of sectors such as the paper industry, where
even minor alterations in products require major changes in large-scale
machinery. This study should thus be considered one step in the emerging
research on the management of technological discontinuities in firms.

Another implication of this study for this field of research is that although the
problems related to managing technological discontinuities are largely
universal, managerial solutions reflect specific national settings. Previous
research on the effects of national context and societal effects on organisations
and technology has found that although the technical core of manufacturing is
often similar, the organisational forms surrounding this technical core vary
across cultures (Child 1981; Lincoln 1990).[2] These cross-national differences
include both cultural aspects such as values (Hofstede 1980) and the structural
characteristics of business firms. For example, corporate strategies (e.g.
Schneider and McDaniel 1991) like patterns of diversification (Kogut *et al.*
1997) as well as the typical corporate hierarchical structures (e.g. Mayer and
Whittington 1996) partly reflect their national context.

Consequently, national specificities are in many ways reflected in how technologies are actually managed and how the technological infrastructure of
corporations is created. The Finnish paper industry, which has been the focus of
this study, represents the typical characteristics of Finnish industries. For
example, Lilja and Tainio (1996: 159) have argued that the typical Finnish firm is
a capital-intensive raw material processor building its competitive position on
an aggressive pattern of investment for economies of scale and upgrading the
system of production. The paper industry companies represent the ultimate end
in this development as they are characterised by technological advancement and
the large scale of their production facilities as an outcome of numerous major
technological change projects. This is to say that although they largely share the
same products and basic technologies as others, Finnish paper industry companies have developed their technologies in a distinctive way and as a result have
become world technological leaders.

One can take two different perspectives on the specificities of the Finnish
paper industry. First, an institutional perspective (e.g. Fligstein 1990; Barley and
Kunda 1992) would claim that such distinguishing content and forms of
economic activities reflect the institutional characteristics of the country in
question. For example, the fact that compared with British industries German
industries are more characterised by quality-oriented craftmanship than by
large-scale batch production, is partly related to the strong presence of a practically oriented tier of higher education in Germany (Sorge 1991). Accordingly,
we can argue that the leading position of the Finnish paper industry is a consequence of several institutional features such as the availability of capital to
convert Finnish forests to exportable products and characteristics of the national
educational system. For example, the fact that engineers are highly valued and
traditionally well-positioned in Finnish industrial companies when compared

for example with the situation in Britain (e.g. Lam 1996) partly explains why Finnish industries are typically technology-driven. Second, in the spirit of recent management studies (e.g. Fondas and Wiersema 1997; Geletkanycz 1997), we can also argue that the dominance of engineers in Finnish paper industry companies has created socialisation mechanisms which keep technological advancement among the most highly regarded managerial objectives.

Thus an institutionalist perspective largely explains the technological strengths of the Finnish paper industry and especially how the industry maintains its high technological standard. However, it can be argued that in order to understand how such technological trends and trajectories were originally created (cf. Karnoe and Garud 1997) we need more specified analyses. For example, it seems that for some reason the paper industry in Finland has been able to create a heroic position comparable to that of Silicon Valley in the United States. As an anecdotal indication of this, working in the major technological change projects in the Finnish paper industry companies in many ways resembles the excellence cultures of the US software companies (see e.g. Schumacher 1997). So it seems that discontinuous technological change is an important trigger for the mobilisation of managerial actors in diverse settings. However, the managerial processes involved here remain largely unclarified. In further studies we therefore need to go deeper into the patterns of meaning among managerial actors in order to understand how technological discontinuities are managed in different industrial and national contexts.

Some practical considerations

In this section we take up some practical considerations, although we believe that it is always difficult and often impossible to give concrete advice to practitioners for managing technological discontinuities. Instead, we believe that research is valuable for practitioners especially when it stimulates their thinking on current work experiences. In this spirit, we see that there are at least three conclusions to be drawn on the basis of the work reported above.

First, this study has confirmed the significance of managerial subdivisions for the management of technological discontinuities. Managers are divided because they represent different professions, take responsibility for a diverse range of managerial functions and have varied backgrounds. Managerial actors are therefore divided both in the ways they act and even more substantially in their approaches to how individual businesses should be developed. Although this notion has pervaded some parts of management literature (see e.g. Knights and Murray 1994 for a review), we believe that individual managers most often do not learn from this phenomenon before they actually attempt to mobilise change in their field of responsibility. This is not to say that these subdivisions would necessarily be harmful, because both the top-down and bottom-up type of organisational change processes have been found to benefit from the diversity within management teams.

We have shown above that diversity within managerial hierarchies becomes especially viable in technological discontinuities. However, instead of offering any practical solutions to decrease this diversity, we conform to some recent writings (e.g. Carlisle and Baden-Fuller 1997), arguing that learning to orchestrate and synchronise this diversity is one of the most critical managerial competences in the current world of corporations. Although it is impossible to define the prescriptive content for such orchestration, we believe that all efforts which increase the amount of dialogue and interaction between conflicting management groupings are of major importance. In order to see each others' points of view, individual managers need interaction and common experiences. Top managers need to scan the views and action of their subordinates and middle managers need to create contacts with people from the upper echelons. Executive involvement is important because otherwise the inherent conflicts within management and organisation which emerge especially when the company is faced with a technological discontinuity cannot be anticipated. Even more importantly, every executive effort to soften formal hierarchical boundaries encourages actors at the lower levels of the managerial hierarchy to initiate and implement new ventures. Managing by walking around is therefore not just a popular slogan but also an important mechanism for both conveying and collecting information throughout managerial hierarchies.

Moreover, though the focus in this study has been on the managers, the issues of mobilisation are also applicable and relevant on the shop floor. The principles of lean organisation (e.g. Womack *et al.* 1990) and projects crossing national boundaries (see e.g. Laurila and Gyursanszky 1998) make the commitment of shop floor level personnel increasingly important. For example, co-operative forms of action between managers and the rest of the personnel may significantly assist the implementation of major technological change projects.

The second practical lesson that can be drawn from this study is related to the competence thresholds inherent in technological discontinuities. We demonstrated above that in addition to posing concrete technical problems, technological discontinuities also pose significant problems for managerial competence and commitment. More concretely, new managerial competence is needed in both the formulation and implementation of major technological change projects. Technological discontinuities by definition involve something not previously done in a corporation. Although a company might have accomplished several successful projects, a new one is always at least partly novel for the individual managers involved. Consequently, technological discontinuities partly invalidate at least those managers who do not upgrade their competences in the process of planning and building the new technologies. Moreover, those managers who are committed to the existing technologies are supposedly not the most willing to initiate adoption of totally new kinds of technology. As a result, because only some actors can be granted the formal responsibility on managing the actual projects, it can be expected that a new frontier is created between those managers who are becoming experts on the new technology and those who are not.

Finally, on the basis of the evidence here it seems that practitioners should be warned about the tendencies of repetition in managing technological disconti-nuities. On the one hand, this means that because technological discontinuities can be considered an ambiguous and uncertain issue, actors most probably rely on their previous experience in managing them. The creativity and innovative-ness of the managers involved should therefore be systematically encouraged and rewarded. On the other hand, the management of technological disconti-nuities can be also repetitious in the sense that such projects are initiated too often. The study above identified long-term patterns of technological develop-ment in the Finnish paper industry context in which major technological change projects appear to have been an almost generalised reaction to various managerial problems. For example, major projects upgrade existing technolo-gies but they may simultaneously create new managerial positions, improve employment in industrial localities and reduce the danger of corporate decline. Consequently, whenever material resources are available, such technological discontinuities in firms are probable because managerial actors have several motives to initiate them. We therefore believe that in technology-intensive industries managers need to take into account that efforts to achieve technolog-ical advancement may sometimes take place at the cost of other equally relevant dimensions in business development.

Notes

1 Introduction

1 It must be noted that after the period examined here Tampella has gone through a transformation in which four out of the company's five business divisions were sold and merged with other companies. Thus the present Tampella – called Tamrock since early 1997 – is not the company examined here.

2 Management and technological discontinuities

1 Paper industry companies usually develop their products by adopting one further capability at a time and use that capability in the upgrading of the existing products.
2 In addition, due to the continuous nature of the production process and the key role of the tacit skills and collective capabilities of employees (see e.g. Penn *et al.* 1992), the success of major technological change projects also depends on the co-operation and commitment of shop-floor workers. For them a successfully managed technological discontinuity produces continuity for both the mill as a workplace and for the mill community as a place to live. Consequently, these actors with their specific objectives and interests also need to be mobilised to facilitate a technological discontinuity.
3 The selection of technological advancement as a major tool for creating competitiveness in the Finnish paper industry has been related to the dominance of engineers in the management of that industry. This dominance of engineers and technical consultants has contributed to the emergence of the current industry recipe, which emphasises the importance of high value-added paper products produced with the most sophisticated production technology available. This recipe was embodied in technological transformation of the industry especially during the 1980s and the early 1990s.
4 As an indication of the recently achieved technological advancement the Finnish paper industry has for several years had the world's most sophisticated large-scale production technology (see e.g. Pekkanen 1989). At the same time, most of the Finnish forest industry corporations which were previously involved in several ranges of business have focused their operations on forest industry products (cf. Räsänen 1993).

3 Mobilisation of managerial actors for discontinuous change

1 This is not to say that the influence of top managers is always positive because personal characteristics such as narcissism may account for various organisational malfunctions (Kets de Vries and Miller 1984; 1985).

2 This social opposition within management may have different forms including open confrontation, compromise, foot-dragging, pretending, working to rule and quitting (see e.g. Weinstein 1979; Lyng and Kurtz 1985; Carnall 1986).

3 In addition, it is possible for the actors to pretend that a change has occurred though it has not (Zald and Berger 1978), or to alienate themselves from the action or to try to exit (Hirschman 1970).

4 The research process

1 In general, the fact that the Finnish paper industry firms have launched so many major technological change projects during recent decades has permitted them to increase their efficiency through project cloning (Lilja *et al.* 1991). Thus the learning curve effect can also be achieved in the repetition of projects with similar features. The cloning approach is also possible between firms when the same managers take part in managing a discontinuity similar to one in which they were already involved in another company.

2 For an overview on the paper industry in the United States see Smith (1997).

3 The Finnish firms have usually not developed new products but instead they have acted as aggressive second-movers on the international paper market with cost efficient variants of better quality paper grades. This process has necessitated capital expenditures amounting to billions of US dollars (see e.g. Paper and Timber 1990).

4 For further discussion on these issues see for example Golden (1997) and Miller *et al.* (1997).

5 Overview of the technological discontinuity in focus

1 All names of individual managers are pseudonyms.

2 MFC grade had been produced by the United Paper Mills at Jämsänkoski, Finland, since 1981. In 1986 and 1987 two other Finnish corporations, Kymmene at Voikkaa and Enso Corporation at Kotka, also started its production though the technology they used was significantly more sophisticated.

6 The actor perspective to the discontinuity: how did management become mobilised?

1 The description deviates from the other actor characterisations in the sense that it does not aggregate and idealise the characteristics of individual managers, but instead relies on secondary sources and the descriptions of other actors.

2 This practice has been common among Finnish paper industry engineers and has tended to equalise production knowledge between the domestic competitors.

3 It must be noted that one of these bankers was in fact a school mate of the current Tampella CEO and had also represented Skopbank on Tampella's board for a year before the take-over.

7 Building a contextual explanatory framework for technological discontinuities

1 Currently, Finnish firms keep these activities separate. In practice, some twenty companies (including Tampella) producing both forest industry machinery and forest industry products have been merged into one paper machine manufacturer and three paper product corporations over the last few decades.

2 The same pattern, although not to the same extent, has been found in the case of the Swedish paper industry corporations (see e.g. Sölvell *et al.* 1991).

3 The merger was technically caused by the fact that the linen mill's production capacity was so huge that it would have gone bankrupt on its own.

4 The background for Tampella's turn to this direction was that at this time knowledge of new innovations could not be easily transferred from abroad and the development work had to be conducted in the existing facilities. During these years Tampella began to plan for machinery using recycled fibres in mechanical pulp production.

5 To mention some details, Tampella's engineering works built the first machine to produce board out of recycled fibre in 1945. Ten years later the use of recycled fibre was further increased after the building of new pulping facilities.

6 In fact, the building of this mill continued the product development of board which had begun during the war years. In 1955 Tampella had already built a packaging facility which converted board into packaging and containers.

7 The use of recycled fibre as raw material in cartonboard had been developed since the war years.

8 This is not to say that the increase in Tampella's financial resources was based solely on changes in corporate ownership. In fact, the evidence presented suggests that it was based on more general factors in the Finnish economy which temporarily altered the conditions for financing industrial corporations. The importance of these conditions is also demonstrated by the fact that a few years later these same owners were unable to help Tampella because their previous operations on the finance markets were frustrated by the international recession, the emerging bank crisis, plummeting real estate and stock prices and the downward spiral of the Finnish economy.

9 In connection with the war indemnities, Finnish engineering works had tried for the first time to build these machines and to develop paper machine designs of their own. The feasibility of this alternative was further promoted by the fact that most of the Finnish engineering works had been involved in the maintenance of Finnish forest industry facilities. These previously acquired capabilities made it easier to convert to paper and board machinery production.

10 The previous deliveries to the Soviet Union were not accepted as references in the West, partly because customers were unable to see the machines in operation.

11 In particular, Valmet was stronger than the other partners because it could deliver machinery for most parts of the paper manufacturing process. Thus its competence increased more rapidly than that of the others, as it could act as a partial supplier in many deliveries which were mainly the responsibility of the others.

12 In 1986 Valmet acquired the Swedish paper machine producer KMW. Although KMW had not delivered board machines for several years and Valmet was still more interested in widening its capabilities in paper machine manufacturing, it did not reject the opportunity to increase its capabilities in manufacturing board machines.

13 As a matter of fact, Valmet's turnover in paper machine manufacturing equalled the turnover of the entire Tampella, of which Tampella's board machinery accounted for less than 10 per cent.

8 Further challenges in the study of managing technological discontinuities

1 Social movement theory was originally developed to explain the emergence and processes involved in spontaneous collective behaviour such as rebellions or religious cults. During recent decades, however, social movement theory has been integrated with general sociological literature on modern society (see e.g. Cohen 1985). This was related to the finding that in many ways social movements resemble conventional organisations and vice versa. Consequently, the social forms which

have been conceptualised as social movements are now diverse. For example, social movements may represent both systematically organised professional activities and collectivist forms of social action (Tilly 1994: 18).

2 It should be acknowledged here that some recent studies have questioned this overall view by presenting evidence which suggests that the characteristics of managerial work differ little between countries (e.g. Lubatkin *et al.* 1995; Markoczy 1997).

References

Ahlfors, M. (1993) Yrityskulttuurin Muutos: Case Tampella Oy, unpublished Master's thesis, Helsinki School of Economics.

Alajoutsijärvi, K. (1996) *A Dyad Made of Steel: Kymmene Corporation and Valmet Paper Machinery and their Relationship, Local Network and Macro Environment 1948–1990*, University of Jyväskylä, Studies 31.

Alho, K. (1961) *Suomen Teollisuuden Suurmiehiä*, Porvoo: WSOY.

Allmendinger, J. and Hackman, J.R. (1996) 'Organizations in changing environments: the case of East German symphony orchestras', *Administrative Science Quarterly* 41: 337–369.

Anderson, P. and Tushman, M. (1990) 'Technological discontinuities and dominant designs: a cyclical model of technological change', *Administrative Science Quarterly* 35: 224–241.

Anthony, P.D. (1986) *The Foundation of Management*, London: Tavistock.

Argyres, N. (1996a) 'Evidence on the role of firm capabilities in vertical integration decisions', *Strategic Management Journal* 17, 2: 129–150.

—— (1996b) 'Capabilities, technological diversification and divisionalization', *Strategic Management Journal* 17: 395–410.

Armstrong, P. (1986) 'Management control strategies and inter-professional competition: the cases of accountancy and personnel management', in D. Knights and H. Willmott (eds) *Managing the Labour Process*, Aldershot: Gower.

Artto, E. (1996) 'Measurement and results of performance and competitiveness of paper industry', *The Finnish Journal of Business Economics* 45, 1: 78–85.

Astley, G.W. and Van de Ven, A.H. (1983) 'Central perspectives and debates in organization theory', *Administrative Science Quarterly* 28: 245–273.

Bantel, K.A. and Jackson, S.E. (1989) 'Top management and innovations in banking: does the composition of the top team make a difference?', *Strategic Management Journal* 10: 107–124.

Barley, S.R. (1986) 'Technology as an occasion for structuring: evidence from observation of CT scanners and the social order of radiology departments', *Administrative Science Quarterly* 31, 1: 78–108.

—— (1990a) 'The alignment of technology and structure through roles and networks', *Administrative Science Quarterly* 35, 1: 61–103.

—— (1990b) 'Images of imaging: notes on doing longitudinal field work', *Organization Science* 1, 3: 220–247.

Barley, S.R. and Kunda, G. (1992) 'Design and devotion: surges of rational and normative ideologies of control in managerial discourse', *Administrative Science Quarterly* 37: 363–399.

Bass, B.M. (1985) *Leadership and Performance Beyond Expectations*, New York: Free Press.

Bessant, J. and Buckingham, J. (1993) 'Innovation and organizational learning: the case of computer-aided production management', *British Journal of Management* 4, 4: 219–234.

Beyer, J.M. and Trice, H.M. (1978) *Implementing Change*, New York: Free Press.

Bierly, P. and Spender, J.-C. (1995) 'Culture and high reliability organizations: the case of the nuclear submarine', *Journal of Management* 21, 4: 639–656.

Bird, B.J. (1989) *Entrepreneurial Behavior*, Glenview, IL: Scott, Foresman.

Björklund, N. (1982) *Kakkosmies: Metalliteollisuutemme Vaiheita Henkilökohtaisesti Koettuna*, Helsinki: Otava.

Bloor, G. and Dawson, P. (1994) 'Understanding professional culture in organizational context', *Organization Studies* 15, 2: 275–295.

Boeker, W. (1989) 'The development and institutionalization of subunit power in organizations', *Administrative Science Quarterly* 34: 388–410.

—— (1992) 'Power and managerial dismissal: scapegoating at the top', *Administrative Science Quarterly* 37: 400–421.

—— (1997) 'Executive migration and strategic change: the effect of top manager movement on product-market entry', *Administrative Science Quarterly* 42: 213–236.

Boisot, M. (1995) 'Is your firm a creative destroyer? Competitive learning and knowledge flows in the technological strategies of firms', *Research Policy* 24: 489–506.

Bower, J.L. (1970) *Managing the Resource Allocation Process*, Boston, MA: Division of Research, Harvard Business School.

Buchanan, D.A. (1993) 'Recruitment mode as a factor effecting informant response in organizational research', *Journal of Management Studies* 30, 2: 297–313.

Burawoy, M. and Hendley, K. (1992) 'Between perestroika and privatisation: divided strategies and political crisis in a Soviet enterprise', *Soviet Studies* 44, 3: 371–402.

Burgelman, R.A. (1983) 'A process model of internal corporate venturing in the diversified major firm', *Administrative Science Quarterly* 28, 2: 223–244.

—— (1988) 'Strategy making as a social learning process: the case of internal corporate venturing', *Interfaces* 18, 3: 74–85.

—— (1991) 'Intraorganizational ecology of strategy making and organizational adaptation: theory and field research', *Organization Science* 2, 3: 239–262.

—— (1994) 'Fading memories: a process theory of strategic business exit in dynamic environments', *Administrative Science Quarterly* 39, 1: 24–56.

—— (1996) 'A process model of strategic business exit: implications for an evolutionary perspective on strategy', *Strategic Management Journal* 17: 193–214.

Burgelman, R.A. and Rosenbloom R.S. (1989) 'Technology strategy: an evolutionary process perspective', in R.A. Burgelman and R.S. Rosenbloom (eds) *Research on Technological Innovation, Management and Policy*, 4: 1–23.

Burgelman, R.A. and Sayles, L.R. (1986) *Inside Corporate Innovation: Strategy, Structure and Managerial Skills*, New York: Free Press.

Burkhardt, M.E. and Brass, D.J. (1990) 'Changing patterns or patterns of change: the effects of a change in technology on social network structure and power', *Administrative Science Quarterly* 35, 1: 104–27.

Burns, T. and Stalker, G.M. (1961) *The Management of Innovation*, London: Tavistock.

Cannella Jr, A.A. and Monroe, M.J. (1997) 'Contrasting perspectives on strategic leaders: toward a more realistic view of top managers', *Journal of Management* 23, 3: 213–237.

Carlisle, Y. and Baden-Fuller, C. (1997) 'Making organizational change work: reshaping beliefs', Paper presented at the Academy of Management Conference, Boston 8–13 August 1997.

Carnall, C.A. (1986) 'Toward a theory for the evaluation of organizational change', *Human Relations* 39, 8: 745–766.

Carpenter, M.A. (1997) 'Putting the upper echelons into context: uncertainty as a moderator of demographic effects', Paper presented at the Academy of Management Conference, Boston 8–13 August 1997.

Castanias, F. and Helfat, C. (1991) 'Managerial resources and rents', *Journal of Management* 17: 151–171.

Chatterjee, S. and Wernerfelt, B. (1991) 'The link between resources and type of diversification: theory and evidence', *Strategic Management Journal* 12, 1: 33–48.

Chen, R. (1996) 'Technological expansion: the interaction between diversification strategy and organizational capability', *Journal of Management Studies* 33, 5: 649–666.

Child, J. (1972) 'Organization structure, environment and performance: the role of strategic choice', *Sociology* 6, 1: 2–22.

—— (1981) 'Culture, contingency and capitalism in the cross-national study of organizations', *Research in Organizational Behavior* 3: 303–36.

—— (1987) 'Managerial strategies, new technology and the labor process', in J.M. Pennings and A. Buitendam (eds) *New Technology as Organizational Innovation*, Cambridge, MA: Ballinger.

—— (1988) 'On organizations in their sectors', *Organization Studies* 9, 1: 13–19.

Child, J. and Francis, A. (1977) 'Strategy formulation as a structured process', *International Studies of Management and Organization* 7, 2.

Child, J. and Smith, C. (1987) 'The context and process of organizational transformation: Cadbury Limited in its sector', *Journal of Management Studies* 24, 6: 565–593.

Child, J. and Yuan Lu (1996) 'Institutional constraints on economic reform: the case of investment decisions in China', *Organization Science* 7, 1: 60–77.

Child, J., Ganter, H. and Kieser, A. (1987) 'Technological innovation and organizational conservatism', in J.M. Pennings and A. Buitendam (eds) *New Technology as Organizational Innovation*, Cambridge, MA: Ballinger.

Clark, J. (1995) *Managing Innovation and Change: People, Technology and Strategy*, London: Sage.

Cohen, J. (1985) 'Strategy and identity: new theoretical paradigms and contemporary social movements', *Social Research* 52: 663–716.

Cohen, M., March, J. and Olsen, J. (1972) 'A garbage can model of organizational choice', *Administrative Science Quarterly* 17, 1: 1–25.

Coleman, S. (1996) 'Obstacles and opportunities in access to professional work organizations for long-term fieldwork: the case of Japanese laboratories', *Human Organization* 55: 334–343.

Crozier, M. (1964) *The Bureaucratic Phenomenon*, Chicago, IL: University of Chicago Press.

Cyert, R.M. and March J.G. (1963) *A Behavioral Theory of the Firm*, Englewood Cliffs, NJ: Prentice Hall.

Daft, R.L. (1978) 'A dual core model of organizational innovation', *Academy of Management Journal* 21, 2: 193–210.

Dahler-Larsen, P. (1997) 'Organizational identity as a "crowded category": a case of multiple and quickly shifting "we" typifications', in S.A. Sackmann (ed.) *Cultural Complexity in Organizations: Inherent Contrasts and Contradictions*, Thousand Oaks, CA: Sage.

Dalton, M. (1950) 'Conflicts between staff and line managerial officers', *American Sociological Review* 15: 342–351.

—— (1959) *Men Who Manage*, New York: Wiley.

D'Aveni, R. (1994) *Hypercompetition: Managing the Dynamics of Strategic Maneuvering*, New York: Free Press.

D'Aveni, R. and Ilinitch, A. (1992) 'Complex patterns of vertical integration in the forest products industry: systematic and bankruptcy risks', *Academy of Management Journal* 35: 596–625.

Davis, G.F. and Thompson, T.A. (1994) 'A social movement perspective on corporate control', *Administrative Science Quarterly* 39, 1: 141–173.

Davis, P., Robinson, R., Pearce, J. and Park, S.H. (1992) 'Business unit relatedness and performance: a look at the pulp and paper industry', *Strategic Management Journal* 13: 349–361.

Dawson, P. (1994) *Organizational Change – A Processual Approach*, London: Paul Chapman.

Day, D.L. (1994) 'Raising radicals: different processes for championing innovative corporate ventures', *Organization Science* 5, 2: 148–172.

Dean, J.W. (1987) 'Building the future: the justification process for new technology', in J.M. Pennings and A. Buitendam (eds) *New Technology as Organizational Innovation*, Cambridge, MA: Ballinger.

Denis, J., Langley, A. and Cazale, L. (1996) 'Leadership and strategic change under ambiguity', *Organization Studies* 17, 4: 673–699.

DiMaggio, P. and Powell, W.W. (1983) 'The iron cage revisited: institutional isomorphism and collective rationality in organizational fields', *American Sociological Review* 48: 147–160.

Donnellon, A., Gray, B., and Bougon, M. (1986) 'Communication, meaning, and organized action', *Administrative Science Quarterly* 31, 1: 43–55.

Dosi, G. (1982) 'Technological paradigms and technological trajectories: a suggested interpretation of the determinants and directions of technical change', *Research Policy* 6: 147–162.

Dosi, G., Freeman, C., Nelson, R., Silverberg, G. and Soete, L. (eds) (1988) *Technical Change and Economic Theory*, London: Pinter.

Dutton, J.E. and Duncan, R.B. (1987) 'The creation of momentum for change through the process of strategic issue diagnosis', *Strategic Management Journal* 8: 279–295.

Dutton, J.E. and Jackson, S.E. (1987) 'Categorizing strategic issues: links to organizational action', *Academy of Management Review* 12: 76–90.

Dyer, W.G. and Wilkins, A.L. (1991) 'Better stories, not better constructs, to generate a better theory: a rejoinder to Eisenhardt', *Academy of Management Review* 16, 3: 613–619.

Ehrnrooth, G. (1991) Hågkomster, Helsinki, unpublished manuscript.

Eisenhardt, K.M. (1989) 'Building theories from case study research', *Academy of Management Review* 14, 4: 532–550.

Eisenhardt, K.M. and Bourgeois, L.J. (1988) 'Politics of strategic decision-making in high-velocity environments: toward a midrange theory', *Academy of Management Journal* 31: 737–770.

Eisenhardt, K.M. and Schoonhoven, C.B. (1990) 'Organizational growth: linking founding team, strategy, environment, and growth among U.S. semiconductors ventures, 1978–1988', *Administrative Science Quarterly* 35: 504–529.

Ettlie, J.E., Bridges, W.P. and O'Keefe, R.D. (1984) 'Organization strategy and structural differences for radical versus incremental innovation', *Management Science* 30, 6: 682–695.

Fennell, M.L. (1984) 'Synergy, influence, and information in the adoption of administrative innovations', *Academy of Management Journal* 27, 1: 113–129.

Ferree, M.M. and Miller, F.D. (1985) 'Mobilization and meaning: toward an integration of social psychological and resource perspectives on social movements', *Sociological Inquiry* 55: 38–61.

Field, R. (1989) 'The self-fulfilling prophecy leader: achieving the metharme effect', *Journal of Management Studies* 26, 2: 151–175.

Finkelstein, S. (1992) 'Power in top management teams: dimensions, measurement, and validation', *Academy of Management Journal* 35: 505–538.

Finkelstein, S. and Hambrick, D.C. (1990) 'Top management team tenure and organizational outcomes: the moderating role of managerial discretion', *Administrative Science Quarterly* 35: 484–503.

—— (1996) *Strategic Leadership: Top Executives and Their Effects on Organizations*, Saint Paul, MN: West Publishing Company.

Fligstein, N. (1990) *The Transformation of Corporate Control*, Cambridge, MA: Harvard University Press.

Fondas, N. and Wiersema, M. (1997) 'Changing of the guard: the influence of CEO socialization on strategic change', *Journal of Management Studies* 34, 4: 561–584.

Fulop, L. (1991) 'Middle managers: victims or vanguards of the entrepreneurial movement?', *Journal of Management Studies* 28, 1: 25–44.

Fynes, B. and Ennis, S. (eds) (1997) *Competing from the Periphery: Core Issues in International Business*, Dublin: Dryden Press.

Gagliardi, P. (1986) 'The creation and change of organizational cultures: a conceptual framework', *Organization Studies* 7, 2: 117–134.

Galbraith, J. (1974) 'Organization design: an information processing perspective', *Interfaces* 4: 28–36.

Geletkanycz, M.A. (1997) 'The salience of "culture's consequences": the effect of cultural values on top executive commitment to the status quo', *Strategic Management Journal* 18, 8: 615–634.

Ghemawat, P. (1993) 'The risk of not investing in a recession', *Sloan Management Review* Winter: 51–58.

Ginsberg, A. and Abrahamson, E. (1991) 'Champions of change and strategic shifts: the role of internal and external advocates', *Journal of Management Studies* 28, 2: 173–190.

Glaser, B. and Strauss, A. (1967) *The Discovery of Grounded Theory: Strategies for Qualitative Research*, Chicago, IL: Aldine.

Globerman, S. and Schwindt, R. (1986) 'The organization of vertically related transactions in the Canadian forest products industries', *Journal of Economic Behavior and Organization* 7: 199–212.

Golden, B.R. (1992) 'The past is past – or is it?: the use of retrospective accounts as indicators of past strategy', *Academy of Management Review* 35, 4: 848–860.

—— (1997) 'Further remarks on retrospective accounts in organizational and strategic management research', *Academy of Management Journal* 40, 5: 1,243–1,252.

Golden, K. (1992) 'The individual and organizational culture: strategies for action in highly-ordered contexts', *Journal of Management Studies* 29, 1: 1–21.

Goodstein, J., Gautam, K. and Boeker, W. (1994) 'The effects of board size and diversity on strategic change', *Strategic Management Journal* 15: 241–250.

Goold, M. and Campbell, A. (1987) *Strategies and Styles: The Role of the Centre in Managing Diversified Corporations*, Oxford: Basil Blackwell.

Gordon, G. (1991) 'Industry determinants of organizational culture', *Academy of Management Review* 16, 2: 396–415.

Gouldner, A. W. (1954) *Patterns of Industrial Bureaucracy*, New York: Free Press.

—— (1957) 'Cosmopolitans and locals: toward an analysis of latent social roles – 1', *Administrative Science Quarterly* 2, 3: 281–306.

—— (1958) 'Cosmopolitans and locals: toward an analysis of latent social roles – 2', *Administrative Science Quarterly* 2, 4: 444–480.

Grieco, M. (1988) 'Birth-marked? a critical view on analysing organizational culture', *Human Organization* 47, 1: 84–86.

Grieco, M. and Lilja, K. (1996) 'Contradictory couplings: culture and the synchronisation of opponents', *Organization Studies* 17, 1: 131–137.

Grieco, M. and Whipp, R. (1991) 'Dismantling logics of action: dilemmas in change', *International Studies of Management and Organization* 21, 4: 78–85.

Halme, M. (1997) 'Ympäristön ja liiketoiminnan suhdetta koskevien perususkomusten muutos UPM-Kymmenessä 1990-luvulla', *Liiketaloudellinen Aikakauskirja* 46, 1: 48–62.

Hambrick, D.C. and Mason, P.A. (1984) 'Upper echelons: the organization as a reflection of its top management', *Academy of Management Review* 9, 2: 193–206.

Hambrick, D.C. and Siegel, P.A. (1997) 'Pay dispersion within top management groups: harmful effects on performance of high-technology firms', in L.N. Dosier and J.B. Keys (eds) *Call to Action*, Academy of Management Best Paper Proceedings 1997.

Hamel, G. (1991) 'Competition for competence and inter-partner learning within international strategic alliances', *Strategic Management Journal* Summer Special Issue, 12: 83–103.

Hamel, G. and Prahalad, C.K. (1994) 'Competing for the future', *Harvard Business Review* 72, 4: 122–128.

Harris, S. (1994) 'Organizational culture and individual sensemaking: a schema based perspective', *Organization Science* 5, 3: 309–321.

Hartley, J., Bennington, J. and Binns, P. (1997) 'Researching the roles of internal-change agents in the management of organizational change', *British Journal of Management* 8: 61–73.

Hater, J.J. and Bass, B.M. (1988) 'Supervisors' evaluation and subordinates' perceptions of transformational and transactional leadership', *Journal of Applied Psychology* 73: 695–702.

Hellgren, B. and Melin, L. (1992) 'Business systems, industrial wisdom and corporate strategies', in R. Whitley (ed.) *European Business Systems: Firms and Markets in Their Institutional Context*, London: Sage.

Henderson, R.M. and Clark, K.B. (1990) 'Architectural innovation: the reconfiguration of existing product technologies and the failure of established firms', *Administrative Science Quarterly* 35: 9–30.

Hickson, D.J. (ed.) (1993) *Management in Western Europe. Society, Culture and Organization in Twelve Nations*, Berlin: De Gruyter.

Hill, R.C. and Levenhagen, M. (1995) 'Metaphors and mental models: sensemaking and sensegiving in innovative and entrepreneurial activities', *Journal of Management* 21, 6: 1,057–1,074.

Hirsch, E. (1990) 'Sacrifice for the cause: group processes, recruitment and commitment in a student movement', *American Sociological Review* 55: 243–254.

Hirsch, P.M. (1995) 'Tales from the field: learning from researchers's accounts', in R. Hertz and J.B. Imber (eds) *Studying Elites Using Qualitative Methods*, London: Sage.

Hirschman, A.O. (1970) *Exit, Voice and Loyalty: Responses to Decline in Firms, Organizations and States*, Cambridge, MA: Harvard University Press.

Hofstede, G. (1980) *Culture's Consequences: International Differences in Work-Related Values*, Beverly Hills, CA: Sage.

Hosking, D. (1991) 'Chief executives, organising processes and skill', *European Review of Applied Psychology* 41, 2: 95–103.

Hosking, D. and Fineman, S. (1990) 'Organizing processes', *Journal of Management Studies* 27, 6: 583–604.

Howell, J.M. and Higgins, C.A. (1990) 'Champions of technological innovation', *Administrative Science Quarterly* 35: 317–341.

Hrebiniak, L.G. and Joyce, W.F. (1985) 'Organizational adaptation: strategic choice and environmental determinism', *Administrative Science Quarterly* 30, 3: 336–349.

Jaakko Pöyry (1987) *Paper Machines of the World*, Helsinki: Jaakko Pöyry & Co, Consulting Engineers, 7 September 1987.

Jackall, R. (1988) *Moral Mazes: The World of Corporate Managers*, New York: Oxford University Press.

Jackson, S.E. and Dutton, J.E. (1988) 'Discerning threats and opportunities', *Administrative Science Quarterly* 33: 370–387.

Jermier, J., Slocum, J., Fry, L. and Gaines, J. (1991) 'Organizational subcultures in a soft bureaucracy: resistance behind the myth and facade of an official culture', *Organization Science* 2, 2: 170–194.

Jick, T.D. (1979) 'Mixing qualitative and quantitative methods: triangulation in action', *Administrative Science Quarterly* 24: 602–611.

Kanter, R.M. (1982) 'The middle manager as innovator', *Harvard Business Review* 60, 4: 95–105.

—— (1983) *The Change Masters: Corporate Entrepreneurs at Work*, New York: Simon & Schuster.

—— (1988) 'When a thousand flowers bloom: structural, collective and social conditions for innovation in organization', in B. Staw and L. Cummings (eds) *Research in Organizational Behavior* 10: 169–211.

Kanter, R.M., Stein, B.A. and Jick, T.D. (1992) *The Challenge of Organizational Change*, New York: Free Press.

Karnoe, P. and Garud, R. (1997) 'Path creation and dependence in the Danish wind turbine field', in J. Porac and M. Ventresca (eds) *Social Construction of Industries and Markets*, Oxford: Pergamon.

Kets de Vries, M.F.R. and Miller, D. (1984) 'Neurotic style and organizational pathology', *Strategic Management Journal* 5: 35–55.

—— (1985) 'Narcissism and leadership: an object relations perspective', *Human Relations* 38: 583–601.

Klandermans, B. (1984) 'Mobilization and participation: social-psychological expansions of resource mobilization theory', *American Sociological Review* 49: 583–600.

Klandermans, B. and Tarrow, S. (1988) 'Mobilization into social movements: synthesizing European and American approaches', in B. Klandermans, H. Kriesi and S. Tarrow (eds) 'From structure to action: comparing movement participation across cultures', *International Social Movement Research* 1: 1–38.

Knights, D. and Murray, F. (1994) *Managers Divided: Organization Politics and Information Technology Management*, Chichester: John Wiley & Sons.

Knorr-Cetina, K. (1981) 'The micro-sociological challenge of macro-sociology: towards a reconstruction of social theory and methodology', in K. Knorr-Cetina and A. Cicourel (eds) *Advances in Social Theory and Methodology: Toward an Integration of Micro- and Macro-Sociologies*, London: Routledge.

Kogut, B., Walker, G. and Anand, J. (1997) 'Agency and institutions: organizational form and national divergences in diversification behavior', Paper presented at the Academy of Management Conference, Boston 8–13 August 1997.

Kotter, J.P. (1982) *The General Managers*, New York: Free Press.

Kunda, G. (1992) *Engineering Culture: Control and Commitment in a High-Tech Corporation*, Philadelphia, PA: Temple University Press.

Lam, A. (1996) 'Engineers, management and work organization: a comparative analysis of engineers' work roles in British and Japanese electronics firms', *Journal of Management Studies* 33, 2: 183–212.

Lane, P.J. and Lubatkin, M. (1997) 'Relative absorptive capacity and interorganizational learning', Paper presented at the 13th EGOS Colloquium, Budapest 3–5 July 1997.

Langley, A. and Truax, J. (1994) 'A process study of new technology adoption in smaller manufacturing firms', *Journal of Management Studies* 31, 5: 619–652.

Laurila, J. (1989) *The Evolution of Personnel Functions in the Multi-Divisional Corporation* (In Finnish), Helsinki: Helsinki School of Economics, Series D-119.

—— (1992) *Paperitehtaan Liiketoiminnallinen Muutos*, Helsinki: Helsinki School of Economics, Series B-118.

—— (1995) *Social Movements in Management*, Helsinki: Helsinki School of Economics, Series A-100.

—— (1997a) 'The thin line between advanced and conventional new technology: a case study on paper industry management', *Journal of Management Studies* 34, 2: 219–239.

—— (1997b) 'Discontinuous technological change as a trigger for temporary reconciliation of managerial subcultures: a case study of a Finnish paper industry company', in S. Sackmann (ed.) *Cultural Complexity in Organizations: Inherent Contrasts and Contradictions*, Thousand Oaks, CA: Sage.

—— (1997c) 'Promoting research access and informant rapport in corporate settings: notes from research on a crisis company', *Scandinavian Journal of Management* 13, 4: 407–418.

Laurila, J. and Gyursanszky, K. (1998) 'Intra-organizational impediments to the development of shop-floor competence and commitment: the case of a Finnish-owned greenfield paper mill in East Germany', *The International Journal of Human Resource Management* 9, 2: 259–273.

Lawrence, P. and Lorsch, J. (1967) *Organization and Environment: Managing Differentiation and Integration*, Illinois: Irwin.

Leavy, B. and Wilson, D.C. (1994) *Strategy and Leadership*, London: Routledge.

Lee, D.M. and Allen, T.J. (1982) 'Integrating new technical staff: implications for acquiring new technology', *Management Science* 28: 1,405–1,420.

Leonard-Barton, D. (1990) 'A dual methodology for case studies: synergistic use of longitudinal single site with replicated multiple sites', *Organization Science* 1, 3: 248–266.

—— (1992) 'Core capabilities and core rigidities: a paradox in managing new product development', *Strategic Management Journal* 13: 111–125.

Levinthal, D. and Myatt, J. (1994) 'Co-evolution of capabilities and industry: the evolution of mutual fund processing', *Strategic Management Journal* 15: 45–62.

Liedtka, J. and Rosenblum, J. (1996) 'Shaping conversations', *California Management Review* 39, 1: 141–157.

Lilja, K. and Tainio, R. (1996) 'The nature of the typical Finnish firm', in R. Whitley and P.H. Kristensen (eds) *The Changing European Firm: Limits to Convergence*, London: Routledge.

Lilja, K., Räsänen, K. and Tainio, R. (1991) 'Development of Finnish corporations: paths and recipes', in J. Näsi (ed.) *Arenas of Strategic Thinking*, Helsinki: Foundation of Economic Education.

—— (1992) 'A dominant business recipe: the forest sector in Finland', in R. Whitley (ed.) *European Business Systems: Firms and Markets in their Insitutional Context*, London: Sage.

Lincoln, J.R. (1990) 'Japanese organization and organization theory', *Research in Organizational Behavior* 12: 255–294.

Lodenius, E. (1908) *Tammerfors Linne och Jern-Manufaktur Aktie-Bolag 1856–1906*, Helsingfors.

Lubatkin, M.H., Ndiaye, M. and Vengroff, R. (1995) 'The nature of managerial work in three developing countries: a test of the universalist hypothesis', unpublished paper, University of Connecticut.

Lyng, S.G. and Kurtz, L.R. (1985) 'Bureaucratic insurgency: the Vatican and the crisis of modernism', *Social Forces* 63: 901–922.

MacCarthy, J.D. and Zald, M.N. (1977) 'Resource mobilization and social movements: a partial theory', *American Journal of Sociology* 82, 6: 1,212–1,241.

MacGrath, R.G., MacMillan, I.C. and Tushman, M.L. (1992) 'The role of executive team actions in shaping dominant designs: towards the strategic shaping of technological progress', *Strategic Management Journal* 13, Winter: 137–161.

Maidique, M.A. (1980) 'Entrepreneurs, champions and technological innovation', *Sloan Management Review* 21, Spring: 59–76.

March, J.G. (1962) 'The business firm as a political coalition', *Journal of Politics* 24: 662–678.

Markides, C. and Williamson, P. (1996) 'Corporate diversification and organizational structure: a resource-based view', *Academy of Management Journal* 39, 2: 340–367.

Markoczy, L. (1996) 'Consensus formation during strategic change', Paper presented at the Academy of Management Conference, Cincinnati August 1996.

—— (1997) 'Sources of shared belief: the role of national-cultural background, other managerial characteristics and strategic interest in shaping beliefs', Paper presented at the 13th EGOS Colloquium, Budapest 3–5 July 1997.

Marttila, I. (ed.) (1988) *Anjalan paperitehdas 1938–1988*, Porvoo: WSOY.

Masuch, M. (1985) 'Vicious circles in organizations', *Administrative Science Quarterly* 30, 1: 14–33.

Mayer, M. and Whittington, R. (1996) 'The survival of the European holding company: institutional choice and contingency', in R. Whitley and P.H. Kristensen (eds) *The Changing European Firm: Limits to Convergence*, London: Routledge.

Meek, V.L. (1988) 'Organizational culture: origins and weaknesses', *Organization Studies* 9, 4: 453–473.

Merton, R.K. (1957) 'Patterns of influence: local and cosmopolitan influentials', in *Social Theory and Social Structure*, Glencoe, IL: Free Press.

Meyer, M.W. and Zucker, L.G. (1989) *Permanently Failing Organizations*, Newbury Park, CA: Sage.

Miller, C.C., Cardinal, L.B. and Glick W.H. (1997) 'Retrospective reports in organizational research: a reexamination of recent evidence', *Academy of Management Journal* 40: 189–204.

Miller, D. and Droge, C. (1986) 'Psychological and traditional determinants of structure', *Administrative Science Quarterly* 31: 539–560.

Miller, D. and Friesen, P.H. (1980) 'Momentum and revolution in organizational adaptation', *Academy of Management Journal* 22: 591–614.

Miller, D. and Toulouse, J. (1986) 'Strategy, structure, CEO personality, and performance in small firms', *American Journal of Small Business* 10, 3: 47–62.

Mintzberg, H. (1985) 'The organization as a political arena', *Journal of Management Studies* 22, 2: 133–154.

Mitchell, W. (1989) 'Whether and when? Probability and timing of incumbents' entry into emerging industrial subfields', *Administrative Science Quarterly* 34: 208–230.

Mohr, L. (1982) *Explaining Organizational Behavior*, San Francisco, CA: Jossey-Bass.

Nadler, D.A., Shaw, R.B., Walton, A.E. and Associates (1997) *Discontinuous Change. Leading Organizational Transformation*, San Fransisco, CA: Jossey-Bass.

Nelson, R.R. and Winter S.G. (1982) *An Evolutionary Theory of Economic Change*, Cambridge, MA: Harvard University Press.

Noble, D.F. (1984) *Forces of Production: A Social History of Industrial Automation*, New York: Alfred A. Knopfl.

Noon, M. (1994) 'From apathy to alacrity: managers and new technology in provincial newspapers', *Journal of Management Studies* 31, 1: 19–32.

Nutt, P.C. (1986) 'Decision style and strategic decisions of top executives', *Technological Forecasting and Social Change* 30: 39–62.

O'Reilly, C.A. III, Caldwell, D. and Barnett, W. (1989) 'Work group demography, social integration, and turnover', *Administrative Science Quarterly* 34: 21–37.

Papadakis, V. and Bourantas, D. (1997) 'The CEO as corporate champion of technological innovation: an empirical investigation', A paper presented at the Academy of Management Conference, Boston 8–13 August 1997.

Paper and Timber (1990) 'The Finnish forest industry: an overview', *Paper and Timber* 72, 2: 86–98.

Parker, M. (1995) 'Working together, working apart: management culture in a manufacturing firm', *Sociological Review* 43, 3: 518–547.

Parker, M. and Dent, M. (1996) 'Managers, doctors, and culture: changing an English health district', *Administration & Society* 28, 3: 335–361.

Pavitt, K. (1984) 'Sectoral patterns of technical change', *Research Policy* 13: 343–373.

—— (1991) 'Key characteristics of the large innovating firm', *British Journal of Management* 2, 1: 41–50.

Pekkanen, M. (1989) 'Structural change in the Finnish forest industry', *Paper and Timber* 71, 2: 104–108.

Penn, R., Lilja, K. and Scattergood, H. (1992) 'Flexibility and employment patterns in the contemporary paper industry: a comparative analysis of mills in Britain and Finland', *Industrial Relations Journal* 23, 3: 214–223.

Pennings, J.M. (1988) 'Information technology in production organizations', *International Studies of Management and Organization* 17, 4: 68–89.

Pennings, J., Barkema, H. and Douma, S. (1994) 'Organizational learning and diversification', *Academy of Management Journal* 37, 3: 608–640.

Pennings, J. and Gresov, C. (1986) 'Technoeconomic and structural correlates of organizational culture: an integrative framework', *Organization Studies* 7, 4: 317–344.

Perrow, C. (1967) 'A framework for the comparative analysis of organizations', *American Sociological Review* 32: 194–208.

Pettigrew, A.M. (1972) 'Information control as a source of power', *Sociology* 6: 187–204.

—— (1973) *The Politics of Organizational Decision Making*, London: Tavistock.

—— (1985) *The Awakening Giant: Continuity and Change in Imperial Chemical Industries*, Oxford: Basil Blackwell.

—— (1987) 'Context and action in the transformation of the firm', *Journal of Management Studies* 24, 6: 649–670.

—— (1992) 'On studying managerial elites', *Strategic Management Journal* 13: 163–182.

Pettigrew, A.M. and MacNulty, T. (1995) 'Power and influence in and around the boardroom', *Human Relations* 48: 845–873.

Pettigrew, A.M. and Whipp, R. (1991) *Managing Change for Competitive Success*, Oxford: Basil Blackwell.

Pizzorno, A. (1985) 'On the rationality of democratic choice', *Telos* 63: 41–69.

Pohjola, M. (1996) *Tehoton Pääoma: Uusi Näkökulma Taloutemme Ongelmiin*, Porvoo: WSOY.

Porac, J.C. and Baden-Fuller, C. (1989) 'Competitive groups as cognitive communities: the case of Scottish knitwear manufacturers', *Journal of Management Studies* 26: 397–496.

Preece, D. (1995) *Organizations and Technical Change. Strategy, Objectives and Involvement*, London: Routledge.

Quinn, J.B. (1979) 'Technological innovation, entrepreneurship and strategy', *Sloan Management Review* 20, Spring: 19–30.

—— (1980) *Strategies for Change: Logical Incrementalism*, Homewood, IL: Irwin.

Räsänen, K. (1993) 'Paths of corporate change and the metaphor of "Sectoral Roots": the case of Finnish corporations 1973–1985', *World Futures* 37: 111–127.

Reicher, S.D. (1984) 'The St Pauls's riot: an explanation of the limits of crowd action in terms of social identity model', *European Journal of Social Psychology* 14: 1–21.

Rogers, E.M. (1962) *Diffusion of Innovations*, New York: Free Press.

Rohweder, T. (1993) *Product Reorientation in the Finnish Paper Industry: Process and Performance*, Helsinki: Helsinki School of Economics and Business Administration, Series: A-93.

Saari, M. (1992) *Minä, Christopher Wegelius*, Jyväskylä: Gummerus.

Saarinen, J. (1994) *Human Resource Function: Changing Roles, A Case Study on Tampella Power Inc.*, University of Tampere: School of Business Administration, Series A1, Studies 40.

Sackmann, S. (1992) 'Culture and subculture: an analysis of organizational knowledge', *Administrative Science Quarterly* 37, 1: 140–161.

Schneider, S.C. (1989) 'Strategy formulation: the impact of national culture', *Organization Studies* 10, 2: 149–168.

Schneider, S.C. and MacDaniel, R.R. (1991) 'Interpreting and responding to strategic issues: the impact of national culture', *Strategic Management Journal* 12, 4: 307–320.

Schumacher, T. (1997) 'West Coast Camelot: the rise and fall of an organizational culture', in S.A. Sackmann (ed.) *Cultural Complexity in Organizations: Inherent Contrasts and Contradictions*, Thousand Oaks, CA: Sage.

Schwenk, C. (1985) 'The use of participant recollection in modeling organizational decision-making', *Academy of Management Review* 10: 496–503.

Scott, W.R. (1998) *Organizations: Rational, Natural and Open Systems*, 4th edn, Upper Saddle River, NJ: Prentice Hall.

Seppälä, R. (1981) *Born Out of the Rapids: Tampella 125 Years*, Helsinki: Tampella.

Sharma, S. (1997) 'A longitudinal investigation of environmental responsiveness strategies: antecedents and outcomes', in L.N. Dosier and J.B. Keys (eds) *Call to Action*, Academy of Management Best Paper Proceedings 1997.

Simpson, R. (1997) 'Rationalisation, flexibility and the impact of presenteeism on the working lives of senior managers in the UK', Paper presented at the 13th EGOS Colloquium, Budapest 3–5 July 1997.

Smelser, N.J. (1963) *Theory of Collective Behaviour* London: Routledge & Kegan Paul.

Smith, K.G., Smith, K.A., Olian, J.D., Sims, H.P., O'Bannon, D.P. and Scully, J.A. (1994) 'Top management team demography and process: the role of social integration and communication', *Administrative Science Quarterly* 39: 412–438.

Smith, M. (1997) *The U.S. Paper Industry and Sustainable Production: An Argument for Restructuring*, Cambridge, MA: MIT Press.

Snow, D.A., Rochford, B.E., Worden, S.K. and Benford, R.D. (1986) 'Frame alignment processes, micromobilization and movement participation', *American Sociological Review* 51: 464–481.

Snow, D.A., Zurcher, L.A. and Eckland-Olson, S. (1980) 'Social networks and social movements', *American Sociological Review* 45: 787–801.

Soeters, J.L. (1986) 'Excellent companies as social movements', *Journal of Management Studies* 23, 3: 299–312.

Sölvell, Ö., Zander, I., and Porter, M. (1991) *Advantage Sweden*, Stockholm: Norstedts.

Sorge, A. (1991) 'Strategic fit and the societal effect: interpreting cross-national comparisons of technology, organization and human resources', *Organization Studies* 12, 2: 161–190.

Spender, J.C. (1989) *Industry Recipes: The Nature and Sources of Managerial Judgement*, Oxford: Basil Blackwell.

Stjernberg, T. and Philips, A. (1993) 'Organizational innovations in a long-term perspective: legitimacy and souls-of-fire as critical factors of change and viability', *Human Relations* 46, 10: 1193–1219.

Strauss, A. (1978) *Negotiations: Varieties, Contexts, Processes and Social Order*, San Francisco, CA: Jossey-Bass.

Strauss, A., Schatzman, L., Erlich, D., Bucher, R. and Sabshin, M. (1963) 'The hospital and its negotiated order', in E. Friedson (ed.) *The Hospital in Modern Society*, New York: Macmillan.

Stuart, T. and Podolny, J. (1996) 'Local search and the evolution of technological capabilities', *Strategic Management Journal* 17, 1: 21–38.

Sturges, J. (1997) 'What it means to succeed: managers' conceptions of career success', Paper presented at the 13th EGOS Colloquium, Budapest 3–5 July 1997.

Tainio, R., Korhonen, M. and Ollonquist, P. (1989) 'In search of institutional management: the Finnish forest sector case', *International Journal of Sociology and Social Policy* 9, 5–6: 88–119.

Tainio, R., Santalainen, T. and Sarvikivi, K. (1991) 'Organizational change in a bank group: the merger process and its generative mechanisms', *Scandinavian Journal of Management* 7, 3: 191–203.

Taylor, V. (1995) 'Watching for vibes: bringing emotions in the study of feminist organizations', in M.M. Ferree and P.Y. Martin (eds) *Feminist Organizations: Harvest of the New Women's Movement*, Philadelphia, PA: Temple University Press.

Teece, D., Rumelt, R., Dosi, G. and Winter, S. (1994) 'Understanding corporate coherence: theory and evidence', *Journal of Economic Behavior and Organization* 23: 1–34.

Teerisuo, J. (ed.) (1972) *Tampellan Inkeroisten tehtaat 1872 – 1972*, Myllykoski: Tampella.

Teulings, A. (1986) 'Managerial labour processes in organised capitalism: the power of corporate management and the powerlessness of the manager', in D. Knights and H. Wilmott (eds) *Managing the Labour Process*, Aldershot: Gower.

Thomas, J.B. and MacDaniel, R.R. (1990) 'Interpreting strategic issues: effects of strategy and the information-processing structure of top management teams', *Academy of Management Journal* 33, 2: 286–306.

Thomas, J.B., Clark, S.M. and Gioia, D.A. (1993) 'Strategic sense-making and organizational performance: linkages among scanning, interpretation, action, and outcomes', *Academy of Management Journal* 36: 239–270.

Thomas, R.J. (1993) 'Interviewing important people in big companies', *Journal of Contemporary Ethnography* 22: 80–96.

Thompson, J.D. (1967) *Organizations in Action*, New York: MacGraw-Hill.

Tilly, C. (1994) 'Social movements as historically specific clusters of political performances', *Berkeley Journal of Sociology* 38: 1–30.

Toch, H. (1966) *The Social Psychology of Social Movements*, London: Methuen.

Torstendahl, R. (1990) 'Essential properties, strategic aims and historical development: three approaches to theories of professionalism', in M. Burrage and R. Torstendahl (eds) *Professions in Theory and History*, London: Sage.

Touraine, A. (1981) *The Voice and the Eye: an Analysis of Social Movements*, Cambridge: Cambridge University Press.

Trice, H. and Beyer, J. (1986) 'Charisma and its routinization in two social movement organizations', *Research in Organizational Behavior* 8: 113–164.

—— (1991) 'Cultural leadership in organizations', *Organization Science* 2, 2: 149–169.

—— (1993) *The Cultures of Work Organizations*, Englewood Cliffs, NJ: Prentice-Hall.

Tsoukas, H. (1989) 'The validity of idiographic research explanations', *Academy of Management Review* 14, 4: 551–561.

—— (1991) 'The missing link: a transformational view of metaphors in organizational science' *Academy of Management Review* 16, 3: 566–585.

—— (1993) 'Analogical reasoning and knowledge generation in organization theory', *Organization Studies* 14, 3: 323–346.

Turner, R.H. (1981) 'Collective behavior and resource mobilization as approaches to social movements: issues and continuities', *Research in Social Movements, Conflict and Change* 4: 1–24.

Turner, R.H. and Killian, L.M. (1972) *Collective Behavior, 2nd edn,* Englewood Cliffs, NJ: Prentice Hall.

Tushman, M.L. and Anderson, P. (1986) 'Technological discontinuities and organizational environments', *Administrative Science Quarterly* 31, 3: 439–465.

Tushman, M.L. and Rosenkopf, L. (1992) 'Organizational determinants of technological change: towards a sociology of technological evolution', in B. Staw and L. Cummings (eds) *Research in Organizational Behavior* 14: 311–347.

Tushman, M.L. and Romanelli, E. (1985) 'Organizational evolution: a metamorphosis model of convergence and reorientation', *Research in Organizational Behavior* 7: 171–222.

Tyre, M.J. and Orlikowski W.J. (1994) 'Windows of opportunity: temporal patterns of technological adaptation in organizations', *Organization Science* 5, 1: 98–118.

Useem, M. (1995) 'Reaching corporate executives', in R. Hertz and J.B. Imber (eds) *Studying Elites Using Qualitative Methods,* London: Sage.

Vaara, E. (1993) *Finnish-Swedish Mergers and Acquisitions: an Empirical Analysis of Success,* Espoo: Helsinki University of Technology, Industrial Economics and Industrial Psychology: Report No. 147.

Van Maanen, J. and Barley, S. (1984) 'Occupational communities: culture and control in organizations', *Research in Organizational Behavior* 6: 287–365.

Van Maanen, J. and Schein, E.H. (1979) 'Toward a theory of organizational socialization', *Research in Organizational Behavior* 1: 209–264.

von Bonsdorff, L.G. (1956) *Linne och Järn 1–3,* Helsinki: WSOY.

Walton, R.E. (1980) 'Establishing and maintaining high commitment work systems', in J. Kimberly *et al.* (eds) *The Organizational Life-Cycle,* San Francisco, CA: Jossey-Bass.

Watson, T. (1994) *In Search of Management,* London: Routledge.

Webb, J. and Dawson, P. (1991) 'Measure for measure: strategic change in an electronic instruments corporation', *Journal of Management Studies* 28, 2: 191–206.

Weber, M. (1924, 1947) *The Theory of Social and Economic Organization,* trans. T. Parsons, New York: Free Press.

Weick, K.E. (1979) *The Social Psychology of Organizing,* 2nd edn, Reading, MA: Addison-Wesley.

Weinstein, D. (1979) *Bureaucratic Opposition: Challenging Abuses in the Workplace,* New York: Pergamon.

Weiss, A.R. and Birnbaum, P.H. (1989) 'Technological infrastructure and the implementation of technological strategies', *Management Science* 35, 8: 1014–1026.

Westley, F. (1990) 'Middle managers and strategy', *Strategic Management Journal* 11: 337–351.

Westley, F. and Mintzberg, H. (1989) 'Visionary leadership and strategic management', *Strategic Management Journal* 10: 17–32.

Westphal, J.D. (1997) 'For every action, a reaction: how CEOs deal with the loss of power in CEO/board relationships', in L.N. Dosier and J.B. Keys (eds) *Call to Action,* Academy of Management Best Paper Proceedings 1997.

Westphal, J.D. and Zajac, E.J. (1995) 'Who shall govern? CEO/board power, demographic similarity, and new director selection', *Administrative Science Quarterly* 40, 1: 60–83.

Whipp, R. and Clark, P. (1986) *Innovation and the Auto Industry: Product, Process and Work Organisation*, London: Frances Pinter.

Whitley, R. (ed.) (1992) *European Business Systems: Firms and Markets in Their Institutional Context*, London: Sage.

Whitley, R. and Kristensen, P.H. (eds) (1996) *The Changing European Firm: Limits to Convergence*, London: Routledge.

Whittier, N. (1997) 'Political generations, micro-cohorts, and the transformation of social movements', *American Sociological Review* 62: 760–778.

Wiersema, M.F. and Bantel, K.A. (1992) 'Top management team demography and corporate strategic change', *Academy of Management Journal* 35, 1: 91–121.

Willmott, H. (1987) 'Studying managerial work: a critique and a proposal', *Journal of Management Studies* 24, 3: 249–270.

Wilson, D.C., Hickson, D.J. and Miller, S.J. (1997) 'The implementation of strategic change: patterns of successful decision making in the face of environmental change', Paper presented at the 13th EGOS Colloquium, Budapest 3–5 July 1997.

Witte, E. (1977) 'Power and innovation: a two center theory', *International Studies of Management and Organization* 7, 1: 47–70.

Wolfe, R.A. (1994) 'Organizational innovation: review, critique and suggested research directions', *Journal of Management Studies* 31, 3: 405–431.

Womack J., Jones, D. and Roos, D. (1990) *The Machine that Changed the World*, New York: Rawson.

Woodward, J. (1965) *Industrial Organization: Theory and Practice*, New York: Oxford University Press.

Yin, R.K. (1989) *Case Study Research: Design and Methods*, London: Sage.

Young, E. (1989) 'On the naming of the rose: interests and multiple meanings as elements of organisational culture', *Organization Studies* 10, 2: 87–206.

Yukl, G. and Van Fleet, D.D. (1992) 'Theory and research on leadership in organizations', in M.D. Dunette and L.M. Hough (eds) *Handbook of Industrial and Organizational Psychology*, 3: 147–197, Palo Alto, CA: Consulting Psychologists Press.

Zald, M.N. (1970) 'Political economy: a framework for comparative analysis', in M.N. Zald (ed.) *Power in Organizations*, Nashville, TN: Vanderbildt University Press.

Zald, M.N. and Berger M.A. (1978) 'Social movements in organizations: coup d'état, bureaucratic insurgency and mass movement', *American Journal of Sociology* 83, 4: 823–861.

Zald, M.N. and Useem, B. (1987) 'Movement and countermovement interaction: mobilization, tactics and state involvement', in M.N. Zald and J.D. MacCarthy (eds) *Social Movements in Organizational Society*, New Brunswick, NJ: Transaction.

Zaltman, G., Duncan, R. and Holbek J. (1973) *Innovations and Organizations*, New York: Wiley.

Index

Printed in the United States
by Baker & Taylor Publisher Services